The Hidden Factor

An Approach For Resolving Paradoxes of Science, Cosmology and Universal Reality

By

Avtar Singh

© 2003 by Avtar Singh. All rights reserved.

No part of this book may be reproduced, stored in a retrieval system, or transmitted by any means, electronic, mechanical, photocopying, recording, or otherwise, without written permission from the author.

ISBN: 1-4033-9363-X (Paperback)
ISBN: 1-4033-9364-8 (Hardcover)

Library of Congress Control Number: 2002095702

This book is printed on acid free paper.

Printed in the United States of America
Bloomington, IN

1stBooks - rev. 07/03/03

Acknowledgement

The author wishes to acknowledge his wife Sumeet Singh and sons, Randeep Singh and Mandeep Singh, for their patience and understanding throughout the duration of this work. The author expresses thanks to many friends, colleagues, and members of family and community for their valuable dialogue and sharing of experiences that contributed to this book. The author is indebted to all the authors and researchers whose publications are referenced in the present work for providing the information and knowledge that were crucial in the accomplishment of this work.

Table of Contents

Acknowledgement iii
Preface ix
Chapter 1 1
Introduction
   Purpose
   Motivation for This Work
   Scientific Method and Spontaneity in Nature 10
Chapter 2
A Modified Specific Theory of Relativity (MSTR) 18
   A Review of Einstein's Specific Theory of Relativity (ESTR)
   Postulates 21
   Speed of Light 22
   Proposed Modification to the Second Postulate of ESTR ... 24
   Time Dilation and Effective Speed of Light 29
   Mass-Energy Behavior in Einstein's Specific Theory of
   Relativity (ESTR) 37
Chapter 3
Gravity Nullification Model (GNM) 41
   Gravity Nullification Model (GNM) 42
   Dynamics of a self-decaying versus non-decaying mass 51
   Spontaneity and Coherence Parameters 57
   Connectivity or Non-locality Explained by GNM 60
Chapter 4
Wave-Particle Duality Based on the Gravity Nullification Model
.... 63
   GNM Based Wave-particle Duality Model 65
   Comparison of de Broglie and GNM Based Wave-particle
   Models 68
   Visibility of an Entity as a Rigid Body 77
   Effect of Rest Mass on Frequency and Wavelength 85
   Wave-particle Behavior of a Photon During Absorption and
   Emission 90
   Wave-particle Duality and Non-locality 102
Chapter 5
A Universe Model based on the Gravity Nullification Model . 104

The Big Bang Model (BBM) 104
Shortfalls of the Big Bang Model (BBM) 107
GNM Based Model of the Universe 115
Results of GNM Based Model of the Universe 128
GNM resolves the Shortfalls of the Big Bang Model (BBM) 150
Comparison of GNM against Recent Astronomical Observations 160
Comparison of GNM against GÖdel's solution to Einstein's field equations 197
What is the role of Time? Is the universe accelerating? 201
Mystery of Dark Galaxies.. 203

Chapter 6
Unraveling of Quantum Mechanics Using the Gravity Nullification Model 206
   Heisenberg's Uncertainty.. 207
   'What is a Quantum Particle?' - Physical Limits of Quantum Behavior 216
   Quantum Paradoxes... 228
   Theory of Parallel Universes Explained by GNM............... 248

Chapter 7
Evaluation of Quantum and Classical Effects of Gravity using the Gravity Nullification Model... 251
   Derivation of Planck's Length and Planck's Mass using GNM 259
   Anti-gravity versus Gravity.. 260
   Resolving Black Hole Controversies.................................. 264
   Gravitational Collapse of a Bose-Einstein Condensate 268
   Matter versus Anti-matter 273

Chapter 8
Summary and Conclusions of the Gravity Nullification Model 276
   Speed of Light 277
   Mass-energy Behavior and Spontaneous Decay of Mass or Particles... 280
   Wave-particle Duality and Spontaneous Decay of Mass or Particles... 281
   GNM Resolves Shortcomings of BBM............................... 283
   Heisenberg's Uncertainty (or a Newtonian mind-set) 284

GNM Resolves Quantum Paradoxes 285
Quantum Versus Classical Gravity 286
Chapter 9
What does it all Mean? 288
Scientific Reality Versus Existence 288
   A Perspective on Time and Evolution 288
   A Perspective on Scientific Reality 293
   A Perspective on Science and Religion 296
   Paul Davies [39], providing an eloquent perspective on science and religion, states: 296
   The Law of Certainty and Free Will 305
References 307

Preface

This work is to put forward an approach for resolution of the paradoxes and questions that haunt modern science and cosmology. It also provides a fresh perspective on scientific reality as it relates to the ultimate universal reality. A critical review of the current scientific method and theories is undertaken with an objective to facilitate their enhancement towards integrating purpose and meaning. I am not an established physicist, chemist, astrophysicist, astronomer or a biologist. Four years ago, I could never imagine publishing the proposed theory that could possibly explain the unresolved questions or paradoxes of modern science. I still do not believe that it is I who is responsible for the theoretical work described in this book. How and what led me to carryout this work and for what reason is still a great mystery to me. It has never been a conscious decision on my part to do what has been accomplished via or through me. Some mysterious power or inspiration has pulled me into and thru this. What is written in these mathematical derivations is like a poem uttered by a mystical poet expressing his innate but extraordinary and miraculous experiences happening inside the emptiness of his being.

What I can recall, it all started one evening about four years ago while helping my son with his high school physics homework. As I was walking away after giving him some hints to solve a problem, my son asked me- "Dad, does physics have anything to do with the real life? My teacher says that it does not." My answer was a casual agreement with the teacher due to the common belief that physics deals with the reality of the inanimate matter and not the real life or consciousness. However, his question triggered an anxiety or a deep-rooted question within my subconscious. Deep inside I felt that my answer to his question had no basis, it seemed like a programmed or cultured response by any scientist to such a question. We are taught or trained to think that life runs at its own guided by some unpredictable and mystical laws, while the matter behaves in a deterministic manner subject to the known laws of nature such as the laws of motion, gravity etc. A physicist seldom questions what is behind the animate behavior or free will of living things such as animals or human beings. Spontaneity or consciousness is often characterized as the 'Ghost in the Atom' and considered to be outside the realms of physical sciences. Or, worse yet, it is often assumed that some existing laws predetermine even the animate or conscious behavior and there is nothing like free will that exists.

Since physicists mostly pursue and understand the laws of the inanimate matter, they consciously ignore any reality that may exist beyond the inanimate matter. Sadly, any open discussion or even a reference to a consciousness-based scientific approach is tabooed among everyday professional scientific dialogue. Since, it is beyond the grasp of the modern scientific method, which is geared to investigate and describe the behavior of the inanimate matter only, any new theory or approach involving spontaneity, consciousness or free will is sure to be shot down by orthodox scientists on grounds of violation of the widely accepted classical scientific method requiring verification and validation via classical experiments. What if the consciousness or free will is beyond the grasp of the classical experiments that employ classical or fixed space/time and physical measurements suitable only for the classical objects?

Since, I did not consider myself an orthodox classical scientist, I did not hesitate to venture beyond the fixed boundary walls of the established scientific method, with a hope that it can be further enhanced to provide answers to more important and deeper questions of life and the universe. The science of the inanimate matter has resulted in a universe that is thought to be meaningless and headed towards extinction into the oblivion. The erroneous conclusions of the current incomplete theories of the inanimate matter including the Newtonian mechanics, Big Bang Model (BBM) and particles physics, have misled us to believe that we live in a universe that came into existence for no reason, and which consists of nothing more than a collection of mindless particles moved by blind and purposeless forces towards a pointless final state of nothingness. With such a gloomy state of the perceived scientific reality, there is not much to loose in making an honest and bold attempt to integrate spontaneity or consciousness in the scientific approach. After all, the spontaneity is an inherent and observed reality in nature and can not be ignored forever. In my view, a science that leads to a meaningless universe is a meaningless science. In order to make science more meaningful to life, integration of spontaneity or consciousness into the scientific method can not be avoided. A scientific method devoid of the spontaneity, which is inherent in nature can represent only partial or local reality. Such a method misrepresents the universal or eternal reality and hence, is unable to achieve the so-called Theory of Everything.

My answer to the question put forward by my son continued to disturb me deep inside and I began searching desperately to find an answer if not "the answer". The feeling of vacuum or emptiness inside me kept pushing me to become a seeker of reality and a researcher to dig deeper into physics and philosophy to find some clues to what governs the dynamics of life or the animated matter. Starting from brushing-up my fundamental knowledge of physics, I studied extensively the recent publications dealing with the subject. I began to see some distant parallelism and similarities between the dynamics of inanimate matter and life. A turning point in my understanding was the book- 'The Large, the Small and the Human Mind" by Roger Penrose [33]. I began to believe that the secret of the inner workings of life was hidden in the behavior that governs the physics of the small. I began to focus on developing an understanding of the inner workings of quantum mechanics, which is still a black box to science. Since I wanted to keep my approach simple, I started to look deeper into the specific theory of relativity for possible headways to unravel the mystery of the small and below the quantum scale.

Both quantum mechanics and specific theory of relativity are widely successful theories that have matched the results of multiple experiments. However, a few crucial disagreements remain dealing with the concept of locality versus non-locality and the limitation of the speed of light for any communication between two separated locations in space. In recent experiments, the observed quantum entanglement between two parts of photon carried away to a large spatial separation has been observed to violate the locality or limitation on communication imposed by the speed of light as prescribed by the special theory of relativity. Since classical objects do not experience such an entanglement phenomenon, I realized that the mystery lies somehow in the space-time behavior of small entities or quantum particles. The wave-particle duality is known to be the governing physical phenomenon that dominates the behavior of the small. To my surmise, most experts (particle physicists) focused mostly on the classical behavior of particles in terms of forces and accelerations, ignoring the fundamental fact that the behavior of the small is dominantly wave-like and less of particle-like. As the entity becomes smaller and smaller the classical particle-like behavior becomes relatively extinct. Hence, I was convinced that the classical particle treatment of small entities can not provide physically valid results without incurring singularities or extremely large inaccuracies. Such singularities are apparent when the general theory of relativity is used to simulate a mass confined to a very small point-like region in space (a black hole). Since quantum mechanics is not a physical or mechanistic theory, I began to look harder into the missing physics (the Hidden Factor) in the theory of relativity to predict the behavior of the small without running into singularities.

The classical or Newtonian treatment of the motion of a particle in physics requires external forces to affect changes to the static (zero velocity) or steady state (constant velocity) motion. No changes to the state of motion of a body can occur without an external force being applied to the body. This is basically due to the assumption that matter and particles are inanimate. Following this assumption, there is no scientific basis for existence of any motion in the universe. If the universe is considered to consist entirely of the inanimate matter, which is defined to be devoid of spontaneous motion, there are no existing scientific laws that could explain the existence of the observed motion and evolutionary changes with time in the dynamic universe, such as the observed Hubble expansion. According to the current scientific understanding and widely accepted models of the universe, the source of all motion in the universe is hidden under the assumed Big Bang singularity that presumes to have started it all at the beginning of time ($t = 0$). For science to lay all its trust into a singularity for the source all motion in the universe is more unscientific and "ghostlier" than the belief of a mystic in the hidden or non-material reality. According to the Big Bang theory (with a zero Cosmological Constant), the initial explosion-like jolt from the Big Bang provided all the energy that fuels any observed motion in the universe today. To rely upon a singularity to be the source of all the motion in the universe is obviously a big leap of faith, which scientists consciously ignore in the absence of any other credible theory to rely upon. Many complicated and partial fill-in-the-gap theories, such as the 'super luminous' inflation and the Anthropic principle, are put forward to overcome deficiencies that are inherent in the Big Bang theory in the absence of nothing better to explain the observed behavior of the universe. However, these patch-up or fill-in-the-gaps theories remain unsuccessful in explaining the wide range of observations including the dark matter, mysterious dark energy causing the accelerating expansion of the universe, non-locality and homogeneity in the universe and so on. To top this, the modern physics resorts to the mysticism (a set of unexplained rules) of quantum mechanics to explain parts of the observed behavior of the universe.

The current state of scientific knowledge is a complex web of multiple theories and rules that provide only limited and partial explanations of the observed data. Because of the missing physics of spontaneity or consciousness, the more advanced modern theories such as various versions of the string theory have to resort to an incomprehensible set of multiple dimensions beyond the three spatial dimensions and time comprehensible to human mind. The whole approach seems like a desperate attempt to repair a broken necklace by gluing each pearl individually, rather than running a single seamless thread thru them. Another analogy representing the current state of science and cosmology may be as if the creatures of the 'Flatland' were trying to describe the three dimensional universe. Such an approach is not only uneconomical, but also afflicted by extreme complexity and incomprehensibility even by the accomplished scientists. Science has been progressively breaking reality into smaller and smaller pieces (particles) and when it attempts to put the pieces together, the sum of the parts falls significantly short of the whole.

Science is paying a heavy price for consciously ignoring spontaneity or consciousness, which is the fundamental property of nature and the universe. Science must look beyond the inanimate matter if it has to catch a glimpse of the Theory of Everything. Any theory without the proper inclusion of the inherent spontaneity or consciousness in nature is only a partial theory of the inanimate matter, which science itself has proven to account for only a small fraction of the whole universal reality. The recent observations of an accelerating universe expansion have further shown evidence of a dark or unknown energy in the form of a Cosmological Constant or a vacuum energy inhabiting the universe. Science now has confirmed that the vacuum is not nothing but probably a source of everything including the visible as well as dark matter or energy in the universe. The mystery behind the strange rules or inner workings of the quantum mechanics is hidden behind this unknown energy. Unfortunately, even the most rigorous estimation of this energy by using existing rules of quantum mechanics are calculated to be in error in excess of 100 orders of magnitude from the observed data.

The strength of the scientific method lies in the validation of a proposed theory against the experimental observations. Without the experimental observations, no proof of the exactness of a theory or model is supposed to exist and the theory remains unverified as regards to its capability to represent reality. With so much of the proof being relied upon the observational data, let us look carefully into the process of collecting the observational data in scientific experiments. First of all the data has to be objective, meaning that it cannot be measured by any subjective experience of an observer. This requires an objective measuring device such as a probe or an instrument or a recorder. All such devices are classical devices in the sense that the outcome of their observation or measurement is a fixed or definite quantity rather than a potential probabilistic wave function of quantum mechanics. Hence, even the outcome of an experiment involving a quantum system is a set of classical measurements. Any proof of the exactness of the quantum theory has to rely upon classical measurements, which involve the collapsed wave function rather than the original un-collapsed and probabilistic wave function. It is well known and accepted scientific fact that the collapse of the wave function depends solely upon the consciousness of the observer. Until a conscious human observer looks upon and interprets the measurement, no collapse or a classical measurement is complete and no data exists. Hence, the ultimate quantification of the measurement is not objective but highly subjective to the consciousness of the observer. Hence, it is almost impossible to guarantee a complete absence of subjectivity in any quantum measurement to the consciousness of the observer. However, the degree of subjectivity may vary from low to high depending upon the complexity and degree of the consciousness of the phenomena being observed. If the phenomenon being observed involves only inanimate matter, the degree of subjectivity in observation may be smaller than the subjectivity of observing a conscious phenomenon such as the human behavior. If a consciousness-induced force is causing a phenomenon to occur in the inanimate matter, again the subjectivity of the observation will remain high.

The point of the discussion above is that the subjectivity in any scientific observations, especially the behavior of the small, is unavoidable and the degree of this subjectivity depends directly upon the degree of the consciousness of the observer and the observed phenomenon. This subjectivity results in the uncertainty of measurement or observation. In other words, if we could imagine an ideal observer who has a capability to witness with full consciousness or objectivity what is being observed, the uncertainty due to the subjectivity of observation can be reduced to zero. The important conclusion here is that the uncertainty, such as the Heisenberg's uncertainty, is not inherent in the phenomenon being observed but in the observer or the process of observation itself which is dependent upon the consciousness of the observer or the observed phenomenon. If the consciousness of the observer is enhanced, the uncertainty can be correspondingly reduced and in the limit the true reality can be observed with no uncertainty. The 'Ghost in the Atom' is due to the limited consciousness of the observer. As the observer becomes fully conscious, the 'Ghost' might be realized to be the 'Host in the Atom', which reveals the ultimate scientific or universal reality with all uncertainty dissolved. The very characterization of the spontaneity in nature as a 'Ghost' reflects the serious lack of awareness or conscious ignorance of the fundamental spontaneous force or energy in nature.

It needs to be realized that what is considered to be the objective observation method in science and works well for the classical or inanimate matter phenomenon, is actually highly subjective as regards to the measurement of the quantum and below-quantum-scale phenomena. Unless this barrier of subjectivity is removed, scientific observations will remain devoid of the objectivity that is crucial to its success in achieving the Theory of Everything or in disclosing the ultimate universal reality. The classical reality revealed by science is only a milestone on the path to the ultimate reality that can only be reached via an enhanced observation process that properly integrates consciousness of the observer. The classical reality is not the end of science; it is only the beginning. Some scientists believe that the end of science has been reached and the ultimate theory of everything is almost here. This author believes that the pursuit of the classical reality and adherence to the orthodoxy of the classical method of observation (mistakenly taken to be objective rather than subjective) alone has driven the modern science to a dead end with irresolvable paradoxes and unanswered questions. For the future advancement of science it is necessary to integrate consciousness in the observation process to make it truly objective as regards to the observation of the universal reality rather than the classical or material reality alone. The author refers to such an approach that integrates science with consciousness as the *Scienciousness*.

The approach followed in the work described herein is based on the application of the laws of conservation on a non-local basis. A theory that fails to satisfy the laws of conservation of energy and momentum on a universal basis may provide weird and unexpected results in its prediction of a phenomenon on a local level. Predictions and observations of some quantum phenomenon such as quantum entanglement is one such example. Similarly, quantum mechanics prediction of virtual particles appearing from nothing (vacuum) is another example that appears to violate locally the mass-energy conservation laws. In classical mechanics, it is a widely used practice to use the free body diagram or analysis of a body separated from its environment to analyze the motion of the body. Similar philosophy is employed in the use of inertial frames of reference in Newtonian method, assuming that the inertial frames of references are entirely independent of each other and exist in a fixed space-time. While the inaccuracies induced by these assumptions are minimal at relatively low velocities (V/C <<1), they become excessively large at high velocities experienced by quantum particles such as photons leading to singularities and weird results that do not match observations.

To avoid such problems and inaccuracies, the approach used in the current work is based on a universal energy balance rather than the classical and separated free-body force-displacement model for predicting motions. Keeping track of the energy balance on a universal basis avoids singularities and unexpected or weird predictions lacking a credible physical basis. Moreover, this approach leads to simple, physical and comprehensible solutions that match experiments and observations of the universe. The results also provide a bridge between the Newtonian mechanics, theory of relativity and quantum mechanics explaining their differences as well as supporting their agreements. The spontaneity inherent in the observed wave-particle duality behavior of a quantum particle is well established in scientific theories. The inclusion of a mathematical representation of such spontaneity of a decaying mass (particle) into energy (wave) and vice-versa helps build this missing bridge and fill in the physics inherent in nature but missing in the current theories.

From a philosophical point of view, an important outcome of the current work is to bridge the gap between science and religion via the *Scienciousness*, which integrates scientific method and consciousness. The perceived gap is due to the ignorance of consciousness. In general, science has tried to address only the material reality while the non-material aspects of reality are left to be addressed by religion(s). This partition or duality of the two realities is not inherent in nature, hence the conflict and an unending debate between science and religion goes on. As discussed earlier, science searches for reality via fragmenting the matter into smaller and smaller fragments (particles). Similarly, sectarian religions define the ultimate reality or Truth by fragmenting human consciousness and experiences into multiple sects, rituals, beliefs and moral laws. Approaches of both science and religions are divisionary rather than integrative leading to the conflict that often has led to wars of words or real wars and violence among humanity. A holistic and consciousness-based approach to both science and religion may be useful in resolving the divisionary conflicts between them.

This work is motivated by the observed scientific evidence that the ultimate universal reality is non-local such as the universal laws, which are indivisible and not subject to the censorship or the speed of light barrier. Scientific evidence supports the notion that the universal laws such as the laws of conservation of energy and momentum are eternal and omnipresent in the universe. Hence, the ultimate reality that science and religion are seeking is the one whole universal reality. If science and religion properly address this reality, the two will become harmonious with each other dissolving their apparent conflicts or differences caused by their current pursuit of the fragmented (material versus non-material) reality. The basis for science is knowledge and the basis for religion is belief. Because of this basic difference, the two have not been entirely compatible with each other. Integrating consciousness, which is the basis for universal reality, can eliminate this incompatibility. The ultimate impact of this work will hopefully lead to an integrated approach, the *Scienciousness,* that would replace the current division between science and religion. There would remain no difference between the approaches of a fully conscious scientist and a saint, since both would perceive the same one whole universal reality. Such integration would bring compassion, peace, purpose, and order back into the currently perceived fragmented, chaotic and meaningless universe marred by the endless cycles of time and evolution.

Let me say a few words regarding the venues I explored for a scientific peer review of this work. When I contacted a few institutions of high learning with regard to their possible review of this work, their response was mostly negative for reasons of their institutional policy. Publication in well-known science journals and magazines requires breaking the whole work into small (no more than a few pages) articles or papers due to the limitations imposed on the length of articles accepted for publication. Peer review process then involves a review by a few selected experts in the limited subject matter of the short paper. Hence, the review process requires fragmenting not only the whole theory into pieces, but also the review itself is performed by experts limited to one specific area. Such a fragmentary and limited review process can never do full justice to evaluate the correctness or completeness of this work impacting a wide range of areas of science such as relativity, quantum mechanics, astronomy, cosmology and astrophysics etc. Moreover, the fragmentation into smaller pieces would weaken the demonstrated strength of the holistic approach to predict a wide range of observations ranging from the behavior of small particles to the whole universe. If it were deemed necessary to publish this work in smaller fragments, so that it could become accessible to a larger scientific community, the author would do so with reluctance and extreme caution.

Another important aspect of this work is a potential revolutionary rather than evolutionary change in the way several well-known scientific concepts are understood by classical or orthodox science. Well-known and widely accepted scientific concepts that are subject to new interpretation in this work include- speed of light C, Cosmological Constant, classical versus quantum behavior, uncertainty versus determinism, dark matter versus dark energy, relativity versus quantum mechanics, vacuum versus Zero-point energy etc. A well-trained and orthodox scientist may need extra-ordinary open-mindedness, contemplation and patience in accepting and adapting these new and revolutionary ways to look at the widely established concepts over the history of science.

Finally, the approach recommended by several individual professionals for achieving a proper and balanced peer review was to publish the whole work in its entirety so that it could be made available to a wider community. If any deficiencies are found or questions are raised, they can be addressed in a follow on revision to the initial publication. Due to the significant mathematical content, this version of the publication is aimed for a scientific audience familiar with the basic concepts of the theory of relativity and quantum mechanics, and cosmology.

Chapter 1

Introduction

Purpose

The purpose of this work is to propose an integrated approach to understand basic reality of the small and large entities in the universe. The author would welcome constructive criticism, comments and suggestions by the established experts to enhance credibility, accuracy and usefulness of the proposed theory to enhance understanding of the universal phenomena. The focus is to develop a simple and comprehensible model of the simple and comprehensible universe that can be understood by a common human being with a minimum of confusion and uncertainties. The basic premise of the model is that "God does not play dice with the universe". The implicate order that underlies all the observed phenomena in the universe is explained along with a perspective on uncertainties.

Motivation for This Work

Some scientists may proclaim that the end of science is near and a Theory of Everything is within the reaches of science. This author believes that science has reached a dead-end rather than an end. This dead-end is the direct result of non-inclusion of the inherent spontaneity in nature into the mainstream of scientific theories. Science has yet to address the fundamental mechanism that governs the inner workings of spontaneity or consciousness, which is the reference or basis of all scientific observations and theories. The science of spontaneity, consciously ignored by scientists, is mysteriously buried in the observed uncertainties and probabilities of wave functions of quantum mechanics. The motivation for this work is to enhance the current scientific method via inclusion of spontaneity representing the true and holistic nature of the universe. Such an integrated approach is referred to as the *Scienciousness*. So far a conscious ignorance or taboo of this aspect of reality by science has paralyzed the scientific theories to achieve their potential for revealing the ultimate reality of the universe.

In spite of the great milestones achieved by science during the past few centuries, unresolved questions and unexplained paradoxes remain that threaten to pull the rug from under the picture of reality painted by the combined knowledge of science. Some of these open questions and paradoxes are listed below:

1. Action at a Distance:

Theory of relativity [1] poses a limit on the speed at which a signal can be transmitted from one point to another in the universe. This limit is the speed of light, denoted by the constant C. Quantum mechanics, however, stipulates a connectivity or capability of instantaneous communication in the universe. Such connectivity has been further confirmed theoretically by the famous Bell's Theorem [2] and experimentally demonstrated to be true by experiments such as those carried out by Aspect [2] in the past decade. This creates a big question –

> 'Is there a limit on the maximum speed at which an event can influence another event happening at a different location in the universe?'

2. Nature of Reality [3]:

The human observer observes large objects such as stars, planets and small objects such as rocks, trees and particles.

Some of these appear stationary and some are in motion. The classical mechanics theory describes laws of motion of ordinary small and large objects at relatively lower speeds than the speed of light. Theory of relativity describes the laws of motion at higher speeds close to the speed of light. Quantum mechanics describes the probabilistic behavior of small quantum particles based on Heisenberg's uncertainty principle [1] and wave/particle complementarity principal.

According to the Heisenberg's uncertainty principle, both the location and velocity of a particle can not be specified at any given time without an irreducible uncertainty that exists in nature. So a complete space-time description of a moving particle can not be absolutely specified in terms of deterministic parameters. This raises the following questions-

> 'If a deterministic space-time description of small quantum particles is not possible, how is it possible for a human observer to observe objects that are made of small moving particles, with well defined spatial location and boundaries in real life?'

> 'If the fundamental description of motion of small particles can only be probabilistic, then what makes the laws of classical mechanics so deterministic?

> 'Does God play dice with the universe?'

3. Gravity:

Gravity provides the glue that holds the objects in the universe such as galaxies, stars, and planets together via attractive forces between masses of all sizes. While quantum models of all other forces existing in nature are possible, gravity has defied any quantum description in terms of any fundamental particles. This raises the following question-

'What is the fundamental nature and mechanism of gravity?'

'Does gravity exist as a fundamental force even if there were no mass in the universe? What leads to the force of gravity in the universe?'

4. Missing or Dark Matter in the Universe:

The observed universe is flat requiring that there is sufficient amount of mass in the universe to maintain its density close to the constant value of the critical density in spite of its observed run-away expansion. But the observed amount of mass is significantly insufficient to predict the flat and expanding universe. Predictions from credible theories show that more than ninety percent of the required mass is missing or undetectable through direct observation of the universe. The questions that arise then are-

'Where and in what form is the missing or dark matter [4] in the universe?'

'What are the constituents of the dark matter?'

5. The Big Bang:

Based on the observed Hubble expansion, it is now believed that the universe started with a Big Bang [4] at the beginning of time. This leads to a discontinuity or singularity in the universe model and several questions listed below:

'What caused the Big Bang?'

'What was there before the Big Bang? What were the initial and boundary conditions of the Big Bang?'

'Will the universe collapse again and lead to another Big Bang?'

6. The observed Order, Simplicity and Comprehensibility in the Universe:

To the human observer the universe seems to be extremely orderly and following simple comprehensible laws that are fine-tuned for existence. This leads to the following questions-

> 'What causes the universe to be so orderly and fine tuned?'

> 'Why do we happen to be here at this time to comprehend the simple yet a wonderful and elegant universe? (Anthropic Principle [5])

> 'Why is the universe so comprehensible to human mind?'

7. Number of Universes:

Quantum mechanics' interpretations of the observed universe lead to the possibility of multiple parallel universes [2]. This model of multiple universes explains many of the observed quantum weirdness that can not be explained by a single universe model. But it leads to several questions-

> 'Are there parallel universes?'

> 'If yes, how many and how are they connected?

> 'Do the same laws of nature exist in all universes?'

> 'How are the events in one universe related or connected to the events in the other universes?'

> 'Did all the universes start with the same one Big Bang or individual Big Bangs?'

8. Time:

More recent theories of time [6,7] point to the possibility that there is no real or absolute time in the universe. Theory of relativity states that time is a relative entity that depends upon the velocity of the observer. For an observer moving at the speed of light the clock stops and time completely dilates to zero. The questions that can be asked are-

'Is time a real entity or only a perception of the human mind?'

'If the age of the universe is about fourteen billion years, what is the frame of reference?'

'What happens to time and how it is related in parallel universes?'

'Did time start with the Big Bang? What was the time before the Big Bang? How much time remains until the next Big Bang? What happens to time in a Big Crunch?'

9. The Observer and the Scientific Method:

Science relies on the measurements made by either an instrument or the direct observations made by a human observer to arrive at a scientific conclusion. It is believed that the consciousness of the human observer [2] is what collapses the wave function in quantum mechanics into an observed classical reality. It is also assumed that a human observer is an independent entity unaffected by what is being observed. This leads to the following questions.

'What is the physics behind the phenomena of the collapse of a quantum wave-function?'

'If quantum mechanics stipulates connectivity of the universe, how an observer is independent and unaffected by the observed?'

'Are the observed scientific parameters by classical measuring devices sufficient and adequate for describing the true nature of reality that is?'

'Is the subjective (un-manifested or dark matter) part of reality more dominant than the so-called objective (manifested or material) reality? How do they relate to each other and to the universal reality as a whole?'

'Is the reality observed by science converging or diverging or evolutionary with time?'

With so many of the important questions regarding the fundamental reality in the universe still remaining to be answered, it is premature to declare victory and the end of science [10]. The serious question is whether the classical scientific method, while successful in describing the material part of the universe, has any fundamental weakness that could make it incapable of revealing the true reality of the universe.

The motivation for this work is to accomplish the following:

1. Answer the above questions by reexamining the fundamental assumptions of the scientific method and widely accepted theories of science today.

2. Find the missing physics, which Einstein alluded to as "the Hidden Factor" in the existing theories. This missing link hopefully will close the apparent gaps between the theory of relativity and quantum mechanics, which are two of the most successful theories in their own rights. 'The Hidden Factor' may

also provide answers to many of the open questions listed above.

3. Provide a physical and mechanistic basis to explain an apparent weirdness of quantum mechanics phenomena.

4. Extract some simple comprehensible insights that can benefit a common man to understand reality and to rejuvenate his confidence in science. Science has become too complex, as it requires understanding of complex concepts of uncertainty and probabilities that are too far removed from the grasp of a common man. Science has become a field for the super experts creating a wider gap with the common man. Simplicity and elegance of science needs to be restored to enhance its credibility, comprehensibility and acceptance.

5. Close the gap between science and religion. The ongoing debate between science and religion for the past centuries has been caused by the basic differences in the methods of science and religion to observe and perceive reality. Science observes reality by separating the observer and dividing the observed reality into particles of matter. Religion on the other hand has ignored scientific observations and relies on faith to propagate the concept of a creator- the God. Science has ignored the inherent spontaneity or consciousness in nature, while religion has ignored the actual observations of the universe. An integrated scientific model that includes spontaneity or consciousness will go a long way to answer the fruitless and endless debate between science and religion. The belief is that there is one common reality, which both science and religion attempt to describe in their own frames of references. This work is an attempt to build a bridge of

consciousness between the two apparently different islands of science and religion.

Avtar Singh

Scientific Method and Spontaneity in Nature

Spontaneity is inherent in nature. It is often considered to be a living or paranormal phenomenon beyond the reaches of science. However, spontaneous decay of atoms or particles is well within the realm of science. Quantum mechanics predicts existence of elementary particles that pop in and out of the so-called vacuum or the empty space. This happens too quickly to enable any physical detection by scientific measurements. Science has not as yet fully developed understanding of the fundamental phenomena behind the spontaneous decay of particles, which is still considered a statistical event involving uncertainty. Scientific method has mostly focused its investigations only to the study of the behavior of the inanimate matter, which is non-spontaneous. However, science must realize that it is setting its own limits by pursuing a path that is bounded or constrained by inanimate matter alone. These limits are evidenced by the existing singularities, irresolvable paradoxes and unexplainable observations that haunt even the most successful theories of science today. Dark matter, black hole singularity, accelerated expansion of the universe, weirdness of quantum mechanics, non-locality or apparent unbounded speed of light, almost limitless vacuum energy, and quantum gravity are just a few examples of such unresolved problems or paradoxes. Science has been as adamant in sticking to its method of object-based or matter-based reality as religion has been to its faith-based pursuit of the truth.

While the classical scientific method has been considered to be widely successful in its material pursuits, it has run into immovable barriers prohibiting any further progress towards uncovering the ultimate universal reality. Past successes, in a way, have become barriers to the future advancement of science. The complexity, weirdness and irresolvable paradoxes of scientific theories will not go away, and may perhaps increase over time just as the entropy, so long as the inherent spontaneity in nature is not properly accounted for. Spontaneity is one of the fundamental, and perhaps the most dominant dimension of reality that can not be ignored if a theory of everything is ever to be achieved. There is a rather heavy price to pay for keeping the spontaneity out from the scientific method. The price is not only the burgeoning complexity, uncertainty, singularities, multiple dimensions of existence (as discovered in the string theories) that are incomprehensible to human experience, but also an unrecoverable loss of meaning and purpose in the universe.

Science has yet to realize that confining its method of investigation to inanimate matter (particles, atoms, molecules etc.) alone is highly subjective rather than objective or realistic. What is ironic that science is getting used to defying its own laws and ignore its own findings to get astray in a mess of chaos and complexity, just like a spider getting entangled in its own web. For example, one of the key scientific finding of modern times is that the amount of observable matter in the universe constitutes a very small fraction (less than 1 to 10 percent) of the total energy in the universe, while 90% or more of the universe exists in the form of un-manifested dark matter or dark energy. If science believes this finding of its own and takes it to heart, it would focus its investigations and spend its valuable resources on investigating the non-matter based reality. Another example of such defiance of a common sense approach is the lack of pursuit of the well-established principle of the wave-particle duality. This principle forms the fundamental basis of the quantum mechanics, which has by far been the most successful theories of science at all times. It is a well-accepted fact that as a particle becomes smaller, it behaves predominantly as a wave (energy) rather than a particle (mass). However, the resources spent on investigating the particle or mass behavior is several orders of magnitude greater than studying the wave behavior in spite of the established knowledge that more than ninety percent of the universal reality at smallest scales exists as waves (energy) and not particles (mass).

Let us look at a few examples wherein the science defies its own laws to reach a weird result and then gets entangled in an effort to resolve it leading to a paradox. The laws of conservation of energy, mass and momentum are well established in that they are never to be violated. However, the concept of "Nothing" or vacuum commonly used by science is in direct defiance of this law. According to the laws of conservation, what exists will always exist in one or the other form, which could vary in mass, energy, space and time. What exists can never be non-existent and in a way is eternal and omnipresent, howsoever its form of existence may vary. Similarly, what does not exist can never exist, i.e. nothing can come out of what does not exist. Still, the concept of nothing or vacuum is commonly used in science to correlate universal observations. It is labeled as a weird and paradoxical happening when this concept runs into difficulty in explaining test data or observations such as the runaway expansion of the universe or rotational velocities of stars in galaxies.

As another example, science relies on the concept of the so-called "Dark Matter", since it sounds like something that can exist, rather than existent energy in space that would be rather hard to justify to our orthodox scientific peers as a viable reality. Ironically, science invents the concept of anti-matter, which gives the connotation of matter that can exist rationally and is capable of nullifying the matter to justify existence of "Nothing". The point here is that we, the innovative scientists have become accustomed to get lost and rely on getting lost once again hoping to reach to the intended destination. We invent a new path by defying the well-established laws of nature, run into a brick wall and reinvent a new even more complex path in the hope to recover the lost path to the destination of reality. The whole process creates much complexity and chaos, which makes us feel even more powerful in recovering from it. We blame the nature for having an inherent uncertainty and complexity, which we are responsible for creating and then reward ourselves with the credit for emergence from it. We glorify this whole process by naming it "Creativity". Since, it all happens over time, we use this fact to give validity and credibility to the concepts of evolution, increasing chaos and complexity in nature.

Another famous example is the Heisenberg's Uncertainty principle. This is the fundamental principle behind quantum mechanics and wave-particle duality. The principle states that there is an irreducible uncertainty in nature that prevents a simultaneous and exact quantification of the spatial location and velocity or momentum of a particle. This principle is the core of the quantum mechanics theory, which is widely believed to be an advanced alternative to the classical Newtonian mechanics of fixed space and time meaning that there exists a universal frame of reference wherein the location and speed of a particle can be quantified in absolute terms. However, this principle itself is based on the assumptions of fixed space and time. The theory of relativity has shown that such a frame of reference is invalid at large velocities commonly experienced by quantum particles to which the uncertainty principle is expected to apply. The point is that the Heisenberg's uncertainty principle is based on the fundamental assumptions of the classical mechanics of fixed space and time. The uncertainty propagated by the principle is an artifact of this fundamental assumption. If space and time were treated as relative and not fixed entities as already demonstrated by the theory of relativity, the uncertainty predicted by the Heisenberg's principle would diminish or dissolve.

The above examples are presented to point out that science creates its own complexity and paradoxes by restricting its realm to the classical frame of reference of the inanimate matter alone. And finally, as an escape of the last resort to overcome these shortcomings, science gives in to some totally unbelievable and incredible concepts such as the concept of 'super-luminous' inflation, parallel universes, time-varying laws, and the Anthropic Principle etc. From the perspective of achieving an overall human fulfillment, the pursuit of the classical scientific approach of the inanimate matter alone may seem like walking in the quicksand of a hot unending desert of paradoxes and singularities. This leads to the feeling of despair, meaninglessness or purposelessness in the universe represented by science and so often expressed by accomplished scientists.

Having looked into some of the weaknesses of the existing scientific method, it would be prudent to discuss how some of these shortcomings can be overcome by following a more conscious approach. This may not only simplify the approach but also provide resolution to many of the current problems and paradoxes of science. Let us come back to the laws of conservation of mass, energy and momentum. The laws of conservation are well established in that they can never be violated by any physical phenomenon. Another common understanding is that these laws are eternal and non-local without any space-time restriction such as the limitation of the fixed speed of light. Another property of nature inherent in these laws is the existing spontaneity that allows transformation from one form to another (for example, from mass to energy and vice-versa), such as observed in the wave-particle duality of light and quantum particles. According to the laws of conservation, what exists will always exist in one or the other form, which could vary in mass, energy, space and time. When all that exists transforms entirely into the form of energy or zero mass, it represents the state of the Zero-point energy. In this state, the time is completely dilated with no clock ticking and all measurable distances in space dilated to zero length. This state of existence would show all measurements of length, time, temperature or other physical properties as zero or non-existent. This is definitely not a state of "Nothing" or vacuum as commonly characterized by a classical scientific observer. In fact, a completely rational description of this state could be that this state represents the absolute reference state wherein the eternal laws of the universe are omnipresent in a uniform space signifying non-locality (infinite speed of light). Note that this conclusion is consistent with the observed non-locality in quantum experiments. This simple and economical approach provides a consistent physical description of the so-called vacuum energy and non-locality without resorting to the complexities of dark matter, multiple and still unknown particles, anti-particles or ghostly weirdness built into the quantum mechanics and associated uncertainty principle which limits only a probabilistic description of reality. There is no need to break up the one and only one miraculously simple, orderly and non-local universe into multiple fragments of particles or parallel universes. This approach, based on basic scientific principles, brings back purpose and meaning by unifying all forms of existence with the universal reality emanating from the one Zero-point energy, which is eternal and omnipresent.

The approach based on conservation laws has been employed in this work to formulate a Gravity Nullification model (GNM) that integrates a modified specific theory of relativity, spontaneity (that allows free transformation of mass, energy, space and time according to the laws of conservation) and classical gravity into one simple model. Gravity Nullification Model (GNM) provides the missing physics in the specific theory of relativity to explain the fundamental relationship between mass and energy, space and time. Some fundamental assumptions in the Einstein's specific theory of relativity are reinterpreted or modified to explain the existing paradoxes of science. GNM consists of a physical model that includes the effects of spontaneous decay of mass and phenomenon of wave-particle duality inherent in nature. Using this model a mathematical relationship is derived relating the wavelength, mass and velocity of a particle as a substitute for the famous de Broglie equation. Observed non-locality and apparent unlimited speed of light is also explained.

GNM is combined with the classical gravitation model and the modified specific theory of relativity to model the universe expansion. This model eliminates singularities in the existing Big Bang models of the universe, explains effects of gravity on the observed mass, dark matter/energy, flatness, accelerated expansion and connectivity (action at distance) in the universe. GNM provides a physical or mechanistic understanding for the existing shortcomings of the Big Bang Model such as the horizon problem (observed uniformity in the universe) and the Cosmological Constant problem without the need for the incredible inflation scenario. A closed form mathematical expression is derived for the Cosmological Constant. Predictions of GNM are compared with the observations of galaxy rotational velocities, luminosity, creation of matter, structure formation and universe expansion etc. The concept of time is clarified with regard to the actual observed universe behavior versus the widely perceived history of the universe evolution predicted by the Big Bang model.

A physical understanding of the inner workings of quantum mechanics is developed using GNM. Heisenberg's uncertainty is revisited and reformulated using the relativistic formulations of GNM. The basis for Heisenberg' uncertainty is shown to be dependent upon the relativistic properties of mass-energy-space-time rather than the widely assumed measurement problem of quantum mechanics. What causes a particle to behave as a quantum versus classical entity is explained by GNM leading to the physical derivations of Planck's scale mass, length and velocities. Paradoxes of quantum mechanics such as the observer paradox (the collapse of the wave-function), non-locality, quantum entanglement, formation and behavior of Bose-Einstein condensates, particle spin etc. are explained using GNM physical models. The theory of parallel universes widely accepted by scientists to explain the inner workings of quantum mechanics is explained in terms of the relativistic formulation of GNM. It is shown that the so-called parallel universes are not exactly parallel (having independent governing laws and parameters) but related to each other via universal laws of conservation of relativistic mass-energy-space-time.

Effects of gravity at quantum and classical scales have been evaluated using GNM, which shows that the widely accepted classical formulation of gravity is consistent with the observed gravitational effects at or below quantum scales. The black hole phenomenon predicted by the general relativity is reevaluated using GNM indicating that a black hole singularity does not exist. A mathematical relationship is derived to predict the gravitational stability radius of a mass irrespective of its scale. Finally, the observed gravitational collapse phenomenon of the Bose-Einstein condensates at large atomic densities is explained using GNM.

The approach proposed in the current work involves some bold new concepts related to phenomena of spontaneity and relativity of mass-energy-space-time. The treatment and results of application of these new concepts are compared against observed behavior of the universe as well as laboratory experiments. The impact of these new concepts span over a wide range of physical phenomena such as gravitation, astronomy, cosmology, and low temperature super-fluidity etc. Because of its multi-disciplinary nature, the author considers this work as a preliminary proposal for review by the scientific community in their various fields of expertise. The author would welcome a serious review of the proposed concepts by the respective experts with a purpose to enhance the holistic understanding of the complex and yet unresolved questions in science. Because of the wide-ranging implications, both scientific and philosophical, the overall results of the work have been integrated in this book to enable a comprehensive and un-fragmented review by a wider community and range of experts. The hope is that this exercise would stimulate the thought process and creativity in resolving the existing paradoxes and in achieving the Theory of Everything.

Chapter 2

A Modified Specific Theory of Relativity (MSTR)

In the Galilean-Newtonian inertial frames of references which are either still or moving at a constant velocity, it is assumed that the lengths of objects and the rate at which time passes remain unchanged from one frame of reference to another. This leads to a different observed velocity of a moving object when observed from one frame of reference versus the other which is moving at a different velocity. The experiments performed by A. A. Michelson and E. W. Morley [1] in the 1880s showed that the speed of light measured in different directions with respect to the Earth's motion, which represented different frames of references, remains constant. Einstein's specific theory of relativity (ESTR) [1,8] forwarded in 1905 resolved this discrepancy. The theory was embodied in two postulates below.

Postulate 1: The laws of physics are the same in all uniformly moving frames of references.

Postulate 2: The speed of light through empty space is invariant regardless the motion of the source or the motion of the observer.

In order to describe motion of a body, we must specify how the body alters its position in space with time. The position is described with respect to a system of coordinates. The Galileian-Newtonian system of coordinates represents those reference frames wherein the laws of the mechanics of Galilei-Newton can be regarded to be valid. Any coordinate system, which is in a condition of uniform motion of translation in a straight line relative to a given Galileian-Newtonian co-ordinate system, is also a Galileian-Newtonian co-ordinate system. According to the principle of relativity, general laws describing the physical description of all natural phenomena remain same in all Galileian-Newtonian frames of references. If all natural phenomena were accurately described by the classical mechanics using Galileian-Newtonian system of coordinates, this principle of relativity would hold valid. But the observed constancy of the speed of light in all frames of references pointed to the inadequacy of the classical mechanics to describe completely all physical phenomena, especially those phenomena that are related to electrodynamics and optics. In spite of this inadequacy, the classical mechanics does predict with great accuracy the motion of heavenly bodies as well as worldly objects encountered in our daily lives. This is fortuitous because the velocities involved in such motion are very much smaller than the speed of light.

Einstein's specific theory of relativity changed the notion of fixed space and time as assumed in the Galileian-Newtonian or classical mechanics. Time, is not absolute and varies according to the speed of the inertial frame. Two events that are simultaneous in one frame are not simultaneous as seen by an observer in the second frame of reference. The result of the specific theory of relativity leads to the following equation for time and space dilation as a function of the speed V of the moving frame of reference for an observer in the fixed (V=0) frame of reference (see Figure 2-1):

$$t = t_o \sqrt{1 - (V/C)^2} \qquad (2\text{-}1)$$

$$S = S_o \sqrt{1 - (V/C)^2} \qquad (2\text{-}2)$$

$$M = \frac{M_o}{\sqrt{1 - (V/C)^2}} \qquad (2\text{-}3)$$

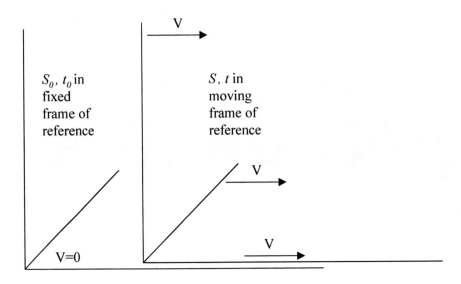

Figure 2-1: Fixed and moving Galileian-Newtonian frames of references.

A Review of Einstein's Specific Theory of Relativity (ESTR) Postulates

In this section we will evaluate the basis and results of the two postulates of the specific theory of relativity proposed by Einstein by comparing them with the observed reality. Specifically, we will take a closer look at the second postulate, which imposes an apparent limit on the action at a distance by comparing it with the recent experimental results. This evaluation will provide us strengths and weaknesses in the postulates that we will try to address later by forwarding new interpretations or models to fill in the gaps what Einstein referred to as "the Hidden Factors".

Postulate 1

The first postulates simply states that the same laws in all frames of references govern the physical phenomena in the universe. That means that there is only one universe and a valid set of physical laws determine the outcome of a phenomenon or event with a given set of initial and boundary conditions. The initial and boundary conditions may vary from event to event or phenomenon to phenomenon but the universal law that governs the progress or outcome remains invariant for any observer or frame of reference from which the observation is made. This postulate also assumes that there exists a fundamental order in the universe that can be described by a set of fixed laws. The varying outcomes are the results of the varying input conditions (initial and boundary conditions) going into these fixed laws. For example, a cup made of glass may break when it falls from a height of ten feet but may not break if it falls from a height of only an inch above the concrete floor. In both cases the same universal law of gravity governs its fall.

This is a very important postulate; if it were not true then the universe would be perceived as a chaos without an order or there may exist multiple universes with varying orders or laws, which is counter to human experience. To a common human observer, the universe is a cosmos with a well-perceived order evidenced by validated scientific and mathematical laws.

Postulate 2

The second postulate states that the speed of light remains constant in any frame of reference irrespective of the speed of the frame. This postulates also leads to the limitation on the maximum speed at which an entity or signal can travel in any frame of reference. This limitation is the speed of light, which also prohibits non-locality and action at a distance in the universe. The basis for this postulate is the results of the experiments such as those by A. A. Michelson and E. W. Morley, which demonstrated the constancy of the speed of light in various frames of references with a varying degree of alignment with the earth's motion and moving at different speeds. This led to the abandonment of the ether existing as a stationary medium in space, which could affect the speed of light depending upon whether its motion opposes or reinforces the propagation speed of the light waves.

Results of some recent experiments have shown that the action at a distance and absence of non-locality are possible, and hence the second postulate of the Einstein's specific theory of relativity (ESTR) may not hold true. These results point to the existence of much higher speeds of light than a constant value of 300,000 kilometers per second in the second postulate. This mounting evidence forces us to reevaluate the basis of the postulate and constancy of the speed of light to explain the observed experimental results.

Speed of Light

Is the speed of light constant at 300,000 kilometers per second? Or, is it higher and possibly infinite as evidenced by the quantum mechanics and related experiments. In order to properly address these questions, let us look into the basic definition of velocity as defined in classical mechanics.

The speed of an object in Newtonian mechanics is defined as the ratio of the elapsed distance over elapsed time. This definition is based on the fundamental assumption that both the elapsed space and time are separately measurable and distinct quantities whose ratio leads to a unique value of the velocity. However, ESTR formulation shows that both space and time dilate as the velocity of the moving frame of reference increases from zero. Hence, space and time are not distinct but relative quantities whose magnitudes are not independent but related directly to the magnitude of the velocity of the frame of reference itself. When V is very much smaller than C, the space and time dilation is negligible and hence an object is physically perceived to be changing location from one spatial position to another in a perceived or measurable elapsed time of the clock. As V approaches a value of C, the size of the object and distance traversed by it dilate to zero and no physical experiment can be performed to directly measure the speed of the object that has vanished or become invisible to the observer in the fixed frame of reference.

There is one other dilemma that challenges our common sense notion of physical reality when we use the concept of velocity in conjunction with the motion of light. As discussed above, when V approaches the value of C, both the elapsed distance and time dilate to zero or the clock stops from moving. The velocity of light, which is the ratio of elapsed distance over time, looses its meaning from mathematical point of view (ratio of zero distance over zero time). Also, from the point of view of physical reality it becomes incomprehensible for a photon of light to move at the speed of light in a zero space with zero time. This provides us a motivation to look into an alternative physical interpretation of the space-time relationship, which provides a description of motion consistent with reality in all frames of references with speeds varying from zero to C.

Proposed Modification to the Second Postulate of ESTR

This author proposes the conservation of elapsed distance and time as a possible alternative to the concept of the speed of light used in the second postulate of ESTR. It is proposed that elapsed distance-time are conserved in relation to each other during the motion of light in the same manner as the mass-energy is conserved during the transformation of mass to energy or vice versa. This is justified due to the following observed relationship between space and time for motion of light in any two different frames of references as shown in Figure 2-1 earlier:

$$\frac{S_0}{t_0} = \frac{S}{t} = C \qquad (2\text{-}4)$$

It is to be noted that the above relationship is the same relationship used in derivation of the Lorentz transformation based on equal speed of light in different frame of references. Hence, the alternative interpretation of space-time conservation during the motion of light does not in any way change the mathematical formulation of Lorentz transformation or ESTR. For this reason, all equations such as equations 2-1, 2-2 and 2-3 as well other equations of ESTR to be discussed later remain equally valid under the new interpretation of the phenomenon of light. The only difference is in the interpretation of C. In the Modified Specific Theory of Relativity (MSTR), C is described as a constant of elapsed distance-time conservation, while in ESTR, C is described as the velocity of light. Later we will show that according to MSTR, the effective speed of light observed in different frames of references moving with varying velocities, by an observer in a fixed frame of reference may vary while the constant of conservation, C, remains invariant.

For a body or entity moving at a velocity less than C in any frame of reference, the space and time are not conserved as described in the following discussion. Let us consider again, as in Figure 2-1, a set of stationary and a second frame of reference that is moving with a velocity V relative to the stationary frame. If a body moves with a velocity W in the moving frame of reference, then

$$\frac{S}{t} = W$$

Lorentz transformation and ESTR provide the following relationship for the velocity W_0 of the same object as observed by an observer in the stationary frame of reference:

$$\frac{S_0}{t_0} = W_0 = \frac{V+W}{1+\frac{VW}{C^2}}$$

From the above it is clear that when W is less than C,

$$\frac{S}{t} \neq \frac{S_0}{t_0}$$

Thus the elapsed distance and time are not conserved from one frame of reference to another. As W equals C, the above relationship converts to the following space-time conservation law for all frames of references as derived earlier:

$$\frac{S}{t} = \frac{S_0}{t_0} = C$$

It should be noted that if the speed V of the moving frame equals C, then irrespective of the value of the object speed W in the moving frame, the above law of space-time conservation holds true. The impact of either V or W approaching the value of C can thus be summarized as follows:

Any object, entity or phenomenon (such as light) moving at speed C in any arbitrary frame of reference, which is moving at a speed V with respect to the stationary frame of reference, obeys the laws of elapsed distance-time conservation. Similarly, any real world object, entity or phenomenon (other than light) moving at speed V in a frame of reference which is moving at speed C relative to the stationary frame of reference obeys the laws of space-time conservation. In both of these cases, the space and time are dilated to zero with regard to any physical aspects or properties related to the object, entity or phenomenon.

The physical consequences of space-time dilation to zero are as follows. All physical aspects of any body, entity or phenomenon in a zero space-time will be fully correlated or connected. In other words, when viewed from the stationary frame of reference, spatial non-locality or action-at-distance will apparently exist since the effective space and hence, distances in the frame of the moving entity are shrunk to zero. Also, simultaneity or temporal non-locality among all physical aspects of the body, entity or phenomenon will exist since clock stops and time dilates to zero.

Note that the elapsed distance-time dilates to zero only in the moving frame of reference with V=C. The space and time in the stationary frame of reference remain independent, intact and undulated. Hence, for an observer in the stationary frame of reference all measurements of space-time will not be observed to be correlated, connected or subject to action-at-distance. However, the physical properties (such as spin, polarization, charm, color etc.) that are inherent properties of the moving body will appear to be fully correlated or connected due to the space-time dilation in its own (or moving) frame of reference. When the velocity of the moving frame V or velocity W of the entity in the moving frame are less than C but greater than zero, the degree of correlation will be less than 100% and depend upon the actual magnitudes of V and W that determine the degree of space-time dilation.

The discussion above leads us to propose the following modification to the ESTR Postulate 2:

Modified Postulate 2:

2a. Elapsed distance-time is conserved during motion of light in empty space.

2b. The maximum rate at which elapsed distance is converted to time when light moves through empty space is invariant regardless the motion of the light source or the motion of the observer.

In the Newtonian mechanics, both space and time are considered fixed and independent absolute entities. For a moving body the space can be measured in terms of the elapsed distance and time can be measured in terms of the amount of the ticking of the clock. Hence, the velocity in a Newtonian frame of reference is also an absolute quantity expressed as a ratio of the elapsed distance to the clock time. In the Einstein's theory of relativity, space and time are not independent and absolute entities but linked together via the constancy of the speed of light in all inertial frames of references. What is being proposed here is an alternate relationship that governs the elapsed distance and time dependence, that is the conservation of elapsed distance and time as relative entities in a similar manner as the mass and energy are conserved. This also allows the empty space in the fixed frame of reference (V=0) as an absolute medium in which a moving body travels from one point to another with a prescribed velocity. The elapsed distance, situated in the empty space, depends upon the velocity V of the moving body and is converted to elapsed time as the body moves through the empty space. This gain in elapsed time slows down the clock that leads to the observed time dilation. However, C is now interpreted as a universal conservation constant that determines the rate at which the elapsed time can be dilated into space and vice versa, instead of as the speed of light in the vacuum.

This interpretation of C is also parallel to the role of C in Einstein's famous mass-energy conservation law:

$$E = C^2 m$$

Wherein, C^2 is interpreted as the rate at which mass (m) can be dilated into energy (E). Such an interpretation of C takes away the restriction on the constancy of the speed of light or its maximum allowed limit, which is consistent with the experimental observations and interpretations of non-locality in quatum mechanics.

Another significance of this new interpretation is that the above two laws of conservation of elapsed distance-time and mass-energy together make distance-time-mass-energy as one continuum for the motion of light via their correspondence to a single universal constant C, as expressed below:

$$\frac{E}{m} = \frac{S^2}{t^2} = C^2 \qquad (2\text{-}5)$$

We will later develop more detailed equations describing the distance-time and mass-energy dilation as an explicit function of the velocity of the entity.

Time Dilation and Effective Speed of Light

Using the proposed modified interpretation of the second postulate of ESTR, the generic equations for the space-time dilation can be derived from the ESTR equations (2-1, 2-2 and 2-3).

Starting from equation (2-1), a series of steps lead to an expression for time dilation, $dt_0 = t - t_0$ with reference to the stationary frame in terms of the speed of the object as follows:

$$t_o = \frac{dt_o}{\sqrt{1-(V/C)^2}-1} \qquad (2\text{-}6)$$

In the stationary frame of reference, time t_o is equal to the elapsed distance S_o divided by the velocity of the object V,

$$t_o = S_o / V \qquad (2\text{-}7)$$

Substituting this in equation (2-6) leads to the following expression for time dilation in reference to the stationary (V=0) inertial frame:

$$dt_0 = S_0 \left(\frac{\sqrt{1-(V/C)^2}-1}{V} \right) \qquad (2\text{-}8)$$

This equation can also be written in a non-dimensional form to define a Time Dilation Factor (TDF) as follows:

$$TDF = \frac{C dt_o}{S_o} = \frac{\sqrt{1-(V/C)^2}-1}{V/C} \qquad (2\text{-}9)$$

A generalized variation of the Time Dilation Factor as a function of V is presented in Figure 2-2. As expected, when the velocity of the object equals C, the Time Dilation Factor equals -1. In other words, at V=C, the time dilation is 100% and the clock stops from ticking. If the elapsed distance is measured in the stationary frame of reference, while the elapsed time is dilated in the photon's own frame of reference (V=C), the measured effective speed of light photon would appear to be infinite. It is to be noted that in all the physical experiments, such as those of Alan Aspect, wherein the entanglement between a pair of photons is observed, the distance between the photons is measured in the stationary or the earth's frame of reference and not in the photon's frame of reference. The definition of speed looses its meaning in V=C frame of reference, since there is no ticking clock or time that exist in this frame of reference. In general, at large values the velocity of a body signifies its ability to convert the elapsed space into time. As discussed earlier, since elapsed space and time become relative rather than absolute entities as velocity increases, the meaning and significance of the Newtonian velocity diminishes. At smaller velocities the time dilation decreases and no time dilation occurs as V reduces to zero in the stationary frame of reference. Hence, at smaller velocities, the Newtonian definition of velocity prevails, since elapsed space and time act as independent and absolute entities.

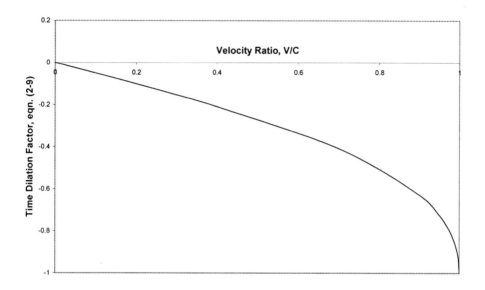

Figure 2-2: Time Dilation Factor versus velocity ratio V/C, equation (2-9).

Another important feature to note here is that the time dilation is directly dependent, according to equation (2-8), upon the elapsed space S_o in the stationary frame of reference. If the motion of the light or the moving object is restricted by an obstruction or boundary, then S_o becomes a finite distance enclosed within the specified boundary and determines the amount of time dilation. The lesser confined is the boundary, the larger is the elapsed space S_o and the time dilation. A more confined boundary of the elapsed distance will reduce S_o and the corresponding time dilation.

Time dilation dt_0 has been expressed above in terms of the elapsed distance in the stationary frame (V=0). In the frame of reference of the light (V=C), time dilates to zero as discussed above. Hence, time in any general frame of reference moving with velocity V is larger or expanded as compared to the frame of reference of the light. The time expansion dt_c relative to the frame of reference of the light can be expressed in terms of V as follows using equations (2-1) and (2-7):

$$dt_c = t = t_o \sqrt{1-(V/C)^2}$$

or,
$$dt_c = t = S_o \frac{\sqrt{1-(V/C)^2}}{V} \qquad (2\text{-}10)$$

Thus the expansion of time in a general frame of reference with a velocity V, relative to the frame of reference of the light is also proportional to the elapsed distance S_o in the stationary frame of reference.

We will now define the concept of the effective velocity, V_{eff}, of an object that is moving with a velocity V_0 in the stationary frame of reference, as the ratio of elapsed distance S_0 to the actual or dilated time in the frame of reference moving with velocity V relative to the stationary frame of reference. As discussed earlier, in all the physical experiments, such as those of Alan Aspect, wherein the entanglement between a pair of photons is observed, the distance between the photons is measured in the stationary or the earth's frame of reference and not in the photon's frame of reference.

$$V_{eff} = \frac{S_o}{t} = V_o\left(\frac{t_o}{t}\right)$$

$$V_{eff} = \frac{V_o}{\sqrt{1-(V/C)^2}} \qquad (2\text{-}11)$$

$$\frac{V_{eff}}{V_o} = \frac{1}{\sqrt{1-(V/C)^2}} \qquad (2\text{-}12)$$

Figure 2-3 shows the variation of V_{eff} versus V. When V is small, V_{eff} equals V_o as expected in the stationary frame of reference. When V approaches C, V_{eff} approaches infinity. This means that if an object were moving at speed of light in the stationary frame of reference, its effective speed in its own frame of reference would become infinity. This is shown in Figure 2-4, wherein C_{eff} represents the effective speed of light in a frame of reference moving at velocity V relative to the fixed frame of reference and C represents speed of light in the fixed frame of reference. As seen in Figure 2-4, C_{eff} remains approximately equal to C until V/C approaches 0.8. As V approaches C, C_{eff} becomes much larger than C approaching infinity at V equal to C. Since, in all the physical experiments, such as those of Alan Aspect [2], wherein the entanglement between a pair of photons is observed, the distance between the photons is measured in the stationary or the earth's frame of reference and not in the photon's frame of reference, the inferred speed of light is same as C_{eff} rather than C. This provides a possible explanation of action at a distance [2] or connectivity in the universe observed in various experiments involving experiments with light [2].

We will discus some more detailed theoretical and experimental validation of the modified Postulate 2 and non-locality aspects of the proposed MSTR in Chapters 3 and 4. We will look into some other fundamental aspects what Einstein may have referred to as "Hidden Variables" or missing physics that may explain the so-called spooky behavior of light photons, electrons and other small quantum particles. In Chapter 3, we will demonstrate that non-locality in the universe results from the fundamental and omnipresent laws of conservation of mass-energy. In Chapter 4, we will show that non-locality can also be explained by the almost infinite wavelength of a particle of negligible mass moving close to the speed of light.

The Hidden Factor

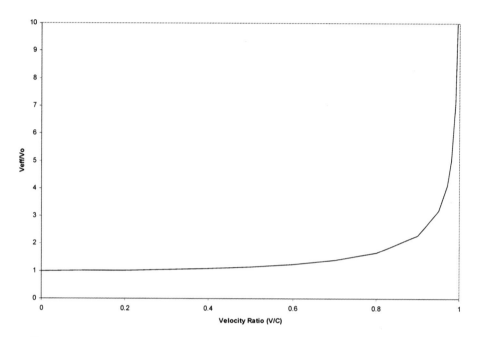

Figure 2-3: Effective velocity, V_{eff}, measured in a frame of reference moving with velocity V relative to the stationary frame.

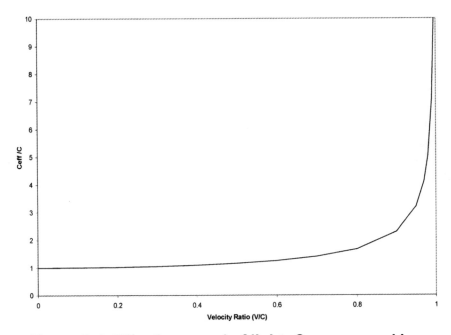

Figure 2-4: Effective speed of light, C_{eff}, measured in a frame of reference moving with velocity V relative to the stationary frame.

Mass-Energy Behavior in Einstein's Specific Theory of Relativity (ESTR)

As described in equation (2-3), the result of the specific theory of relativity leads to the following equation for mass *M* of an object in a frame of reference moving at speed V relative to an observer in the fixed frame of reference (V=0), see Figure 2-1:

$$M = \frac{M_o}{\sqrt{1-(V/C)^2}} \qquad (2\text{-}3)$$

Or,

$$\frac{M}{M_o} = \frac{1}{\sqrt{1-(V/C)^2}} \qquad (2\text{-}13)$$

M_o in the above equation represents the mass of the object in the fixed frame of reference. Figure 2-5 shows how the mass ratio M/M_o in equation (2-13) varies as a function of the velocity ratio V/C. When V equals zero, the mass is equal to the rest mass M_o. As velocity increases, mass increases and when V equals C, M becomes infinitely large. If we imagine several independent inertial frames of references in the universe initially at rest and later acquiring varying non-zero velocities, the total mass-energy of the universe at these two instances will be different. At low velocities (V<<C), the error in mass-energy conservation is small; however, at large velocities (V~C), the error in calculated mass-energy can become infinitely large. Hence, the ESTR formulation does not conserve total or universal mass-energy in relativistic frames of references that are moving at different speeds relative to each other.

Where does this increase in mass-energy come from? Also, such an increase in mass-energy violates the principle of mass-energy conservation. As we discussed earlier, spatial distance dilates to zero at V=C. Thus at V=C, ESTR would lead to an infinite mass confined to a zero volume. Such a scenario provides a singularity in ESTR, which does not seem to represent a common sense physical reality. This leads us to propose a new postulate in the MSTR to account for mass-energy conservation in relativistic frames of references, as described below:

> *Postulate 3:* The total mass-energy (equal to the rest mass-energy, $E_o = M_o C^2$, in the stationary frame of reference) of a system is conserved in all moving frames of references.

Let us consider a system of n number of masses at rest with a total mass of M_o. If later, some or all of the masses acquire motion with non-zero velocities, the resulting kinetic energy of the system must come from conversion of part of the mass M_o to kinetic energy. The net increase in the kinetic energy must equal the net decrease in the total mass of the system. This postulate that is holistic in nature has significant implications on the fundamental understanding we have regarding the everyday reality and phenomena observed in the universe as discussed below and in the following chapters.

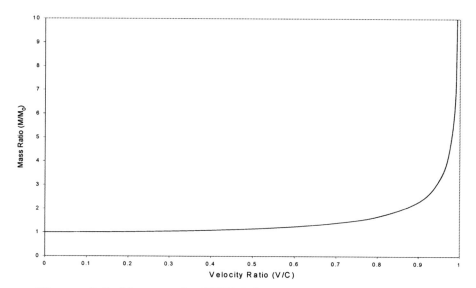

Figure 2-5: Mass ratio (M/M$_0$) for a moving mass versus velocity ratio (V/C).

There is a significant impact of the Postulate 3 on the classical or Newtonian understanding of fixed and independent space and time. Postulate 3 suggests that there are no independent inertial frames of references due to the fact that a finite mass-energy transfer occurs from one frame of reference to another moving at a finite velocity relative to the first, in order to conserve the overall mass-energy of the universe. At low velocities ($V \ll C$) of everyday human experience, the magnitude of such mass-energy transfer is negligible due to the small relativistic effects, hence the assumption of fixed and independent space/time hold approximately true. However, at large velocities close to C, mass-energy transfer as well as space/time dilation are significant leading to a breakdown in the classical assumption of fixed and independent space/time.

In the next chapter, using the Postulate 3 and associated laws of conservation of energy and momentum, we will develop formulation of the Gravity Nullification model to explain the observed behavior of the universe, explain physical basis for the inner workings of quantum mechanics and resolve several of the existing paradoxes of quantum behavior including quantum gravity, entanglement, particle spin and superfluidity etc.

Chapter 3

Gravity Nullification Model (GNM)
The Hidden Factor

In Chapter 2, we focused or discussion on the physical behavior of the space-time in relativistic frames of references with only a minimal discussion on the relativistic mass-energy behavior. We did relate the equivalence between the laws of space-time conservation and the mass-energy conservation expressed in equation (2-5). Let us look deeper into the relativistic mass-energy behavior and explore how it affects not only space-time, but also the fundamental aspects of the observed behavior of light propagation and non-locality.

The famous debate between Albert Einstein and Neils Bohr revolved around three related but distinct questions:

1. Is there an inherent and irreducible uncertainty in prescribing both the position and momentum of quantum entities or particles (The Heisenberg Uncertainty)?
2. Is quantum mechanics a complete theory and does it represent reality?
3. Is non-locality or faster than light communication between two events or phenomena a reality as predicted by the quantum mechanics?

Both Einstein and Bohr provided thought experiments and arguments based on relativistic and quantum mechanics theories. However, Einstein insisted that quantum mechanics is an incomplete theory since it did not provide a deterministic prediction of the reality. His position was – "God does not play dice with the universe" and that every element in a complete physical theory must have a counterpart in the physical reality.

Both the theory of relativity and quantum mechanics successfully predict observed results of several experiments [2,9]. However, at the same time there are a few apparent differences that remain unexplained. These differences relate to the three questions listed above. These questions are so fundamental that the whole foundation of a physical understanding of the nature of reality depends upon satisfactory and accurate answers to these questions. Then there are several as yet unanswered questions in science, which can not be successfully answered until the above questions are resolved. These questions include among others- 'What is Dark Matter and where is it? ', 'Is there a quantum theory of gravity?', 'How is the universe expanding and what is it's future?', and 'Is there a Theory of Everything?' etc.

In this and the following chapters we will look into possible "Hidden Factors" [2] that could fill in the gaps in the physical understanding of the reality and hopefully could help provide answers to the questions above.

Gravity Nullification Model (GNM)

Following the Postulate 3 of the modified specific theory of relativity in chapter 2, the total mass-energy of a system must be conserved in both the moving and the stationary frames of references. Since the total mass-energy in the stationary frame of reference is equal to M_o, the actual mass m in the stationary frame of reference must obey the following relationship so that the total mass-energy in the moving frame remains constant at M_o:

$$m = M_o \sqrt{1-(V/C)^2} \qquad (3\text{-}1)$$

or,

$$\frac{m}{M_o} = \sqrt{1-(V/C)^2} \qquad (3\text{-}2)$$

To check this, let us substitute m from above into equation (2-3). This gives,

$$M = \frac{m}{\sqrt{1-(V/C)^2}} = M_o$$

Equation (3-2) thus provides the law of dilation of the rest mass M_o in the stationary frame of reference to allow a valid moving frame of reference to exist and that conserves mass-energy. Since a moving frame of reference corresponds to a moving body, equation (3-2) provides a necessary condition for motion to occur in a body. In our normal worldly experience, which involves very small velocities (V<<C), the decrease in overall mass of classical moving bodies is very small and not perceptible. The external force exerted on a classical body to cause its motion is caused by the mass dilation, elsewhere in the forcing system, to energy given by equation (3-2). In the theory of relativity, mass-energy conservation rather than formulation of a driving force is used to predict motion of a body.

Since the process of dilation of the mass is opposite to the process of gravitation that causes formation or growth of mass, we refer to equation (3-2) as the Gravity Nullification Model (GNM). We derived GNM above from the Postulate 3 of conservation of mass-energy between different frames of references. We will now demonstrate how GNM can also be derived from the principle of conservation of mass-energy within the body itself.

What is the physical cause of the motion in the universe? In the Newtonian or classical mechanics, a force external to the body causes the motion of a body. The inertia of the body due to its mass opposes the external force acting on the body. The external force comes from an independent entity external to and separate from the body itself. Such a motion is defined as a **non-spontaneous motion**, since it is caused by factors (forces) external to the body.

As part of the ESTR, Einstein derived the famous law of mass-energy conservation:

$$E = C^2 m \qquad (3\text{-}3)$$

Wherein, E and m represent equivalent changes in energy and mass. In the ESTR, a conversion of mass to energy is allowed according to the equation (3-3) above. We can postulate a body that has an inherent capability to convert a part or all of its mass to energy, such as is observed during a spontaneous decay [1] of a particle. In case of elementary particles such as electrons, protons and neutrons this may involve substantial amount of energy as required by $E=mC^2$. Such particles are known to be stable over long times. In contrast, particles of extremely low mass such as photons and multitude of other unstable particles are known to decay instantly [1]. To represent such observed spontaneous decay of particles, the particle theory presumes existence of anti-particle partner for each existing particle, which can be annihilated by the anti-particle and spontaneously converting to energy. Particles that decay instantly can not be easily detected and hence, it is not known as to how many such particles may exist in the universe. It is postulated that the evidence of the existence of dark matter in the universe may point to the possibility that a large fraction of the universe mass-energy may consist of such particles.

The energy released during a spontaneous conversion of mass to energy via a spontaneous decay can be used to provide motion or kinetic energy to the remaining (unconverted) mass of the body or particle. The motion caused via such a postulated process of self-decaying mass that is internal to the body is defined as a spontaneous motion as opposed to the non-spontaneous motion of a classical non-decaying mass or body. Another example of the spontaneous motion could be the Big Bang process that started the dynamic expansion of the universe and which is now believed to be a credible theory of creation and evolution of the universe.

Since the process of spontaneous mass-energy conversion is internal to the body and absence of any energy transfer across a boundary separating the body from its environment, there is no increase in entropy during this process. Hence, the process of spontaneous decay of particles or wave-particle duality is an isentropic and reversible process. On the other hand, motion or kinetic energy of a classical (non-decaying) body caused by an external force involves energy transfer to the body from its environment is an irreversible process leading to an increase in entropy.

Let us now consider a self-decaying mass M_o at rest (V=0). A small portion of the mass then spontaneously converts to energy according to the equation (3-3) above. The converted energy is used by the remaining mass m, to propel itself causing a spontaneous motion with a velocity V. The relativistic kinetic energy of the body with mass m and moving at speed V is given by the following equation from ESTR:

$$KE = mC^2 \left(\frac{1}{\sqrt{1 - \frac{V^2}{C^2}}} - 1 \right) \qquad (3\text{-}4)$$

Equating this kinetic energy to the energy from mass dilation given by equation (3-3), we obtain the following:

$$(M_o - m)C^2 = mC^2 \left(\frac{1}{\sqrt{1 - \frac{V^2}{C^2}}} - 1 \right) \qquad (3\text{-}5)$$

Simplifying the above provides the following equation that is same as GNM represented by equation (3-2):

$$m = M_o\sqrt{1-(V/C)^2} \qquad (3\text{-}2)$$

We have shown that GNM equation (3-2) satisfies the laws of conservation of mass-energy both on a holistic or universal basis as well as within the body itself. It should be noted that when V equals C, the rest mass M_o dilates to zero and converts fully to kinetic energy.

We derived the equation (3-2) of the Gravity Nullification Model using mass-energy balance in equation (3-5). In addition to satisfying the law of conservation of mass-energy, the moving mass m must also satisfy the law of conservation of momentum. This can only be achieved if, during its decay, the body of mass m breaks up into at least two or more (even number) of pieces that move in opposite directions such that the sum of each individual mass-energy equals the total mass-energy, $E_o=M_oC^2$ and the net sum of momentum (mV) of each individual mass equals zero. Figure 3-1 shows a typical simple schematic of such a process, wherein a body of rest mass M_o decays into two smaller masses m_1 and m_2 moving at velocities V_1 AND V_2 respectively. Another example of such motion is shown in Figure 3-2 in which the original mass breaks into four equal masses that move in symmetrically opposite directions to satisfy the momentum conservation law. Note that it is not necessary that the broken masses are all equal to each other. So long as there are at least two masses that move opposite to each other with equal and opposite momentum (mV), there could be different amounts of paired masses that can conserve the momentum. This can be represented mathematically as follows:

$$m V = m_1V_1 + m_2V_2 + m_3V_3 + \ldots\ldots + m_nV_n = \sum_1^n m_nV_n \qquad (3\text{-}6a)$$

In the above equation each term m_nV_n represents a pair of masses moving with equal and opposite momentum relative to the fixed frame of reference. A large number of three-dimensional mass velocity configurations are possible that can satisfy equation (3-6). A two-dimensional representation of one such configuration is shown in Figures 3-2. When V equals C, the mass M_o disintegrates or converts completely into energy.

Also note that the overall mass-energy ($E_o = M_o C^2$) has to be conserved during this decay process such that the following is satisfied:

$$M_o C^2 = \sum_{1}^{n} \frac{m_n C^2}{\sqrt{1 - (V_n / C)^2}} \qquad (3\text{-}6b)$$

It should be noted that under the configurations depicted above, motion at each point in space is directed radially outwards, which is exactly opposite of the direction of gravity force acting on each of the masses. For this reason, as well as those discussed earlier, the model is called Gravity Nullification Model.

The scenario depicted in Figure 3-2 can easily be extended to three dimensions, wherein the mass m_0 gets de-fragmented and dissolved into paired fragments moving radially outwards into space at speed V. It is significant to point out that this three-dimensional model is similar to the Huygens' model of light propagation in that the light is depicted to travel radially outward at each point in space. Huygens proposed that wave fronts of light waves emanating from a point source behave as overlapped crests of secondary waves. This is shown in Figure 3-3. The advancing spherical wave front is an envelope of new overlapping wavelets emanating from all the points along the previous wave front. Only a few of the infinite number of wavelets that combine to give a new smooth wave front are shown in two dimensions in Figure 3-3. Far from the original source, the wave front becomes nearly a plane – as do light waves arriving at the earth from the sun.

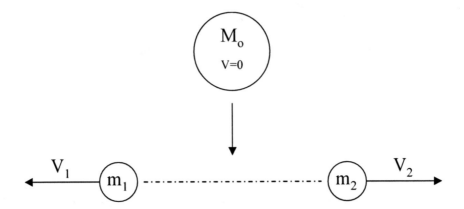

Figure 3-1: Gravity Nullification Model predicted distribution of moving masses during break-up of a self-decaying mass.

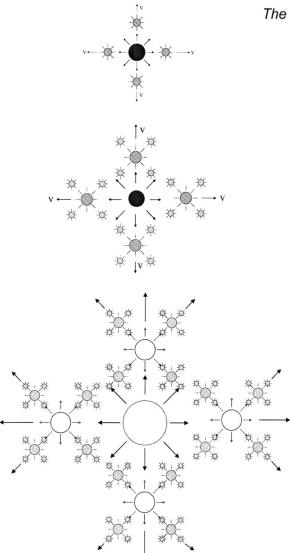

Figure 3-2: Gravity Nullification Model predicted distribution of moving masses during break-up of a self-decaying mass.

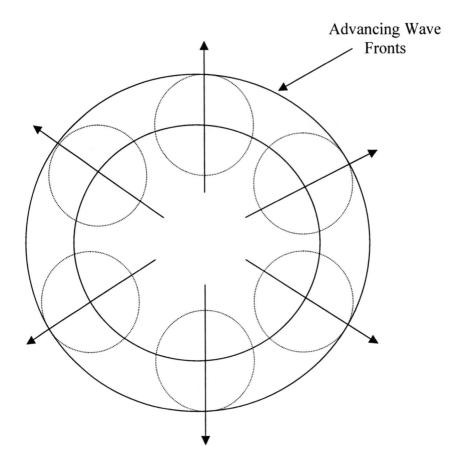

Figure 3-3: Huygens' Model of spherical wave front propagation.

Photons of light emitted by stars in the universe are an ideal example of the physical process described by the Gravity Nullification Model. The mass that converts to energy or light in stars escapes in the form of photons moving at velocity C radially outward in all directions in space. This model can represent the very fabric of the space filled with stars and galaxies in the universe. Later we will develop a detailed model of the universe using this approach to explain the observed characteristics of the universe expansion, dark matter etc.

Dynamics of a self-decaying versus non-decaying mass

In the development of GNM above, we defined the spontaneous motion as the motion caused by a conversion of mass to kinetic energy internal to the body. Such a spontaneous conversion is achieved by a body defined as a **self-decaying mass (SDM)**. Another type of body representing a classical mass observed in our daily world and that does not decay spontaneously, is defined as a **non-decaying mass (NDM)**. In order to understand the behavior of classical bodies versus quantum particles and the universe, it is important to understand the mass-energy characteristics of both types of masses. The kinetic Energy of non-decaying mass (NDM) with a rest mass M_0 is given by equation (3-4):

$$KE_{ndm} = M_0 C^2 \left(\frac{1}{\sqrt{1 - \frac{V^2}{C^2}}} - 1 \right) \qquad (3\text{-}7)$$

The rest mass energy E_0 is represented as follows:

$$E_0 = M_0 C^2 \qquad (3\text{-}8)$$

Combining the above two equations, the following is obtained:

$$\frac{KE_{ndm}}{E_0} = \left(\frac{1}{\sqrt{1-\frac{V^2}{C^2}}} - 1\right) \qquad (3\text{-}9)$$

Similarly, kinetic energy of a self-decaying mass (SDM) with an initial rest mass M_o can be obtained by substituting equation (3-2) into equation (3-4) as follows:

$$\frac{KE_{sdm}}{E_0} = \left(1 - \sqrt{1-\frac{V^2}{C^2}}\right) \qquad (3\text{-}10)$$

Figure 3-4 shows the non-dimensional kinetic energy, equation (3-9), and non-dimensional mass, equation (2-13), for a non-decaying mass as a function of the velocity ratio V/C. Both kinetic energy and mass of a non-decaying mass increase without limit as V increases and approaches the value of C. As noted earlier, this violates the law of mass-energy conservation.

Figure 3-5 shows the non-dimensional kinetic energy, equation (3-10), and non-dimensional mass, equation (3-2), for a self-decaying mass as a function of the velocity ratio V/C. Kinetic energy of a self-decaying mass increases and mass decreases as V increases. As V approaches the value of C, the kinetic energy becomes equal to the rest mass energy while the mass decreases to zero. Thus the mass-energy is fully conserved during the spontaneous motion of a self-decaying mass.

Using equations (2-13) and (3-2), the ratio of the relativistic mass M_{ndm} a non-decaying body to the relativistic mass M_{sdm} of a self-decaying body of equal rest mass, the following can be obtained:

$$\frac{M_{ndm}}{M_{sdm}} = \frac{1}{1-(V/C)^2} \qquad (3\text{-}11)$$

Using equations (2-13) and (3-2) again, the ratio of the rest mass M_{ondm} of a non-decaying body to the rest mass M_{osdm} of a self-decaying body that would result in equal relativistic mass for both at various velocities V, the following can be obtained:

$$\frac{M_{ondm}}{M_{osdm}} = 1-(V/C)^2 \qquad (3\text{-}12)$$

The comparison of the behavior of a non-decaying versus self-decaying mass is shown in Figure 3-6, wherein the ratio of the relativistic masses of a non-decaying body to a self-decaying body of an equal rest mass given by equation (3-11) is plotted as dotted line. At V=0, the ratio is 1, since the rest masses are equal. As V increases, the ratio increases exponentially since the relativistic mass of a non-decaying body increases due to addition of external energy while the relativistic mass of the self-decaying body decreases to provide for the kinetic energy needed for motion. A solid line in Figure 3-6 shows the ratio of the rest mass of a non-decaying body to the rest mass of a self-decaying body that would result in equal relativistic mass for both at various velocities. As V increases above zero, the rest mass of a non-decaying mass has to be smaller than the rest mass of the self-decaying mass to achieve the same relativistic mass to account for the addition of external energy.

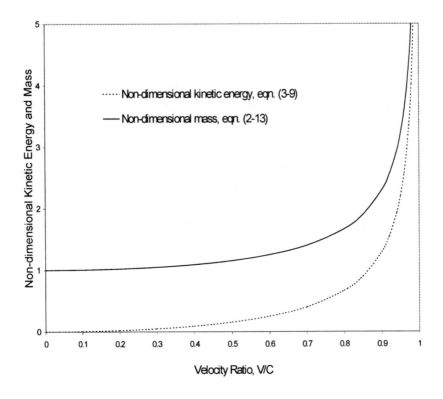

Figure 3-4: Non-dimensional kinetic energy and mass versus velocity of a non-decaying mass.

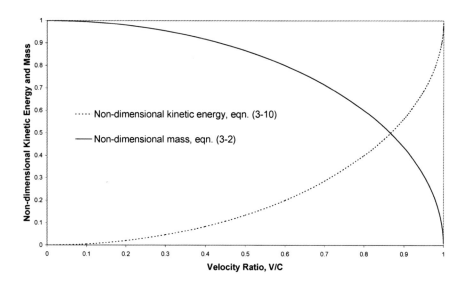

Figure 3-5: Non-dimensional kinetic energy and mass versus velocity of a self-decaying mass.

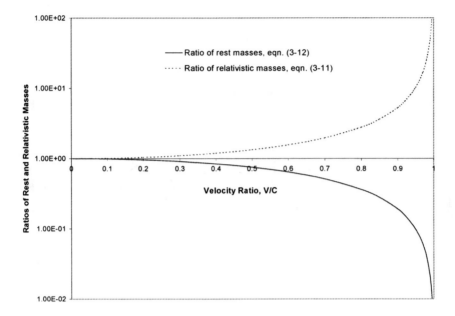

Figure 3-6: Mass ratios for a self-decaying and non-decaying mass.

Spontaneity and Coherence Parameters

In formulation of the specific theory of relativity and GNM equations (3-1) through (3-5), the independent or free parameter is the velocity ratio (V/C) of actual velocity to the distance-time conservation constant C. We define this as the Spontaneity Parameter or the free-will parameter P_{fw}, since it determines the amount of mass, space and time dilation.

$$P_{fw} = \frac{V}{C} \qquad (3\text{-}13)$$

At very small values of the Spontaneity Parameter, the amount of dilation is insignificant and mass, space and time may be treated as fixed or invariant as in the classical Newtonian mechanics. However, when the Spontaneity Parameter is close to unity (V≈C), the dilation is significant leading to various quantum or small scale phenomena counter-intuitive to the classical understanding of the universe.

We will also define a Coherence Parameter, P_{co}, as the angle θ described in terms of the velocity ratio V/C as follows:

$$P_{co} = \vartheta$$

wherein,

$$Cos\,\vartheta = \frac{V}{C} = P_{fw} \qquad (3\text{-}14)$$

and,

$$Sin\,\vartheta = \sqrt{1-(V/C)^2} \qquad (3\text{-}15)$$

This leads to the following form of GNM equation (3-2):

$$\frac{m}{M_o} = Sin\vartheta$$

Figure 3-7 depicts the Coherence Parameter, P_{co}, in a graphical form in terms of velocity vectors V and C. When V is aligned with C, P_{co} =0 and P_{fw} =1.

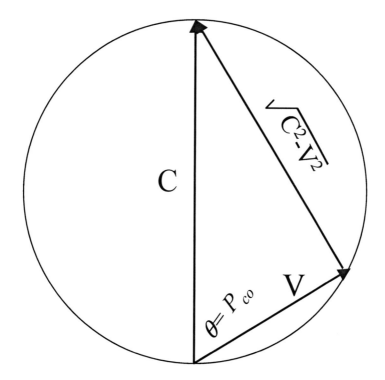

Figure 3-7: Coherence Parameter, P_{co}, in terms of velocity vectors V and C.

Connectivity or Non-locality Explained by GNM

Since the total mass-energy in any frame of reference must be conserved, no frame of reference is independent of any other frame of reference moving with a different speed. It implies that any two frames of references moving relative to each other must transfer mass-energy between them to allow conservation of the total mass-energy. Mass-energy dilation in one frame of reference, as given by equation (3-2), is transferred to the other frame to cause an equal mass-energy expansion such that the total mass-energy of the universe remains constant. Such transfer of mass-energy is possible only through seamless connectivity or non-locality in the universe.

In the Newtonian Mechanics that represents our classical understanding of everyday experience, an isolated object or mass can be accelerated to acquire different peak velocities depending upon the amount of force applied. The object moving at a uniform speed equal to any of the possible values of the peak velocity represents a distinct inertial frame of reference which is assumed to be independent of other frames of references. What is ignored, however, is the fact that the applied force comes from the neighboring frame external to the frame of reference of the moving body and the source of this force is the mass-energy of the neighboring or the connected frame. The connected frame could either be the stationary frame or any other frame moving at a different velocity than the body itself. Because of the absence or lack of consideration for this exchange or connectivity, total or universal mass-energy is not conserved in different Newtonian frames of references and this leads to the common everyday experience or the illusion of the isolated non-locality of the moving body from the rest of the universe.

To clarify this aspect of the Newtonian inertial frames of references, let us consider again the two frames of references in Figure 2-1. Let us now consider a non-moving body in the stationary (V=0) frame of reference. The same body when viewed from the moving frame of reference has a speed that is equal and opposite to the speed V of the moving frame. Hence, the mass-energy of the body in the moving frame is higher than the mass-energy in the stationary frame by the amount of the kinetic energy acquired by the body. Based on the Newtonian laws of motion, this increase in kinetic energy is caused by the action of an isolated force on the body in motion. Such an understanding of the motion observed in everyday experience is so deeply rooted in us that it has become a matter of common sense for us to assume non-locality a foregone reality in the universe. We forget that such an understanding, howsoever approximately true at very low velocities (V<<C), is utterly unscientific since it violates the fundamental law of conservation of mass-energy.

In chapter 2, we demonstrated that the non-locality results from conservation of space-time when the speed V of an entity or its frame of reference equals C. Here we provide a more generic basis for non-locality, which is the conservation of mass-energy in the universe.

Now, let us consider a mass that decays spontaneously according to the Gravity Nullification Model, equation (3-2), and attains a speed V. Because of the increased speed, the elapsed distance and time dilate as per equations (2-1) and (2-2). If the mass completely decays to zero, V becomes equal to C and space and time both dilate to zero. As discussed in Chapter 2, the complete dilation of space and time leads to the observed non-locality and action-at-distance.

It is argued and implied in ESTR that a real signal that carries information in the form of an energy wave with a finite non-zero frequency and wavelength can never travel faster than C. This is consistent with the modified Postulate 3 or the Gravity Nullification Model as follows. The information in a signal is stored in the form of energy waves of specific frequency and wavelength as seen by an observer in the stationary frame of reference. Since the signal maintains a finite energy or non-zero rest mass that constitutes the specific stored information during its transmission through empty space in the stationary frame of refrence, its speed V remains slightly less than (however very close to the value of) C and the space-time does not completely dilate to zero. This preserves the locality aspects of the signal since its velocity V is slightly smaller than C. A photon of light, on the other hand that is not burdened with a signal of a non-zero rest mass, can decay its mass to zero as it travels through the empty space. Hence, non-locality and coherence is observed in the behavior of a photon over the entire universe.

Chapter 4

Wave-Particle Duality Based on the Gravity Nullification Model

In Chapter 3, we investigated the mass-energy behavior in relativistic space and how it affects not only space-time, but also the fundamental aspects of the fabric of the universe in terms of non-locality and action at a distance. We derived the Gravity Nullification Model to explain the observed non-locality in the universe. In this and the following chapters we will address non-locality in terms of the wavelength of a particle as part of the wave-particle duality phenomenon. This will also help us understand the following issues in the famous debate between Albert Einstein and Neils Bohr:

- Is there an inherent and irreducible uncertainty in prescribing both the position and momentum of quantum entities or particles (The Heisenberg Uncertainty)?
- Is quantum mechanics a complete theory and does it represent reality?

The first question above is related to the observed dual behavior of photons and other quantum particles that act both as particles as well as waves. A particle represents a distinct mass or body of matter that can be described quantitatively in terms of its mass, energy, space and time location or motion. A wave on the other hand represents a moving entity of energy spread over a distance measured by its wavelength. Interactions between two particles are by impact and deflection while the waves interact via interference and diffraction. Experiments carried out with photons of light demonstrate that depending upon the kind of the experiment and measurements made, the photon of light can act either as a distinct particle or a wave.

French physicist Louis de Broglie, in his doctoral thesis in 1924, discovered that every particle of matter also acts as a wave that guides its motion in space. Under proper conditions of size and motion of the particle and geometry of interacting bodies (such as slits), every particle or body will produce an interference or diffraction pattern just like a wave. Louis de Broglie proposed the following relationship between the effective wavelength and momentum of the body:

$$Wavelength = \frac{Planck's\ Const.}{Momentum}$$

or,

$$\lambda_{dbr} = \frac{h}{mV} \qquad (4\text{-}1)$$

Where λ_{dbr} is de Broglie wavelength, m is the mass and V is the velocity of the body. A body with a small mass such as electrons will have a large wavelength and exhibit interference and diffraction patterns while passing through small slits or across sharp edges. But a large body such as a canon ball moving at ordinary speeds has a very small wavelength and does not exhibit any wavelike interference and diffraction fringes. It is observed in experiments that the particles travel through the empty space or around obstructions like waves, but are detected as particles when they strike the detector screen. The observed and calculated interference or diffraction patterns of electrons using equation (4-1) match closely, justifying the adequacy of the de Broglie model of wave-particle duality. Even larger particles such as neutrons, protons and whole atoms are observed to exhibit the dual behavior.

GNM Based Wave-particle Duality Model

The model proposed by Louis de Broglie in equation (4-1) was based on classical motion of a body without any consideration of the relativistic effects. The mass of the body was assumed to remain constant irrespective of the magnitude of its velocity. The velocity itself was based on the notion of fixed or non-variant space and time coordinates as in classical Newtonian mechanics. From the relativistic models of GNM presented in Chapter 3, we know that at high speeds relativistic effects become significant and mass-energy-space-time can vary in relation to each other. We will now use the specific equations of GNM to derive a set of relativistic relationships governing the dualistic wave-particle behavior of ordinary bodies. In Chapter 3 we defined two types of masses – self-decaying mass (SDM) and non-decaying mass (NDM). Since different laws govern the mass-energy and corresponding space-time behavior of the two types of masses, we will derive separate mathematical relationships for their wave-particle behavior.

Wave-particle Behavior of a Self-decaying Mass (SDM)

A quantum particle such as a photon is also described as a quantum wave packet with an energy *e* given by the following equation:

$$e = hf \qquad (4\text{-}2)$$

Wherein h is Planck's constant and f is the frequency of oscillation of the wave packet. Now, from ESTR, a mass m has an equivalent energy given as follows,

$$e = mC^2 \qquad (4\text{-}3)$$

Combining the above two equations gives,

$$f = \frac{mC^2}{h} \qquad (4\text{-}4)$$

The wavelength λ of the wave packet is related to the velocity of the wave packet as follows,

$$\lambda = \frac{V}{f} = \frac{hV}{mC^2} \qquad (4\text{-}5)$$

In Chapter 3, we derived the following relationship given by equation (3-2) for a self-decaying mass m with a rest mass M_0 and moving at velocity V,

$$m = M_o \sqrt{1-(V/C)^2} \qquad (4\text{-}6)$$

Substituting the above into equations (4-4) and (4-5), we obtain the following expressions for the frequency f_{sdm} and wavelength λ_{sdm} of the self-decaying mass,

$$f_{sdm} = \frac{M_o C^2 \sqrt{1-(V/C)^2}}{h} \qquad (4\text{-}7)$$

$$\lambda_{sdm} = \frac{hV}{M_o C^2 \sqrt{1-(V/C)^2}} \qquad (4\text{-}8)$$

These equations can also be expressed in the following non-dimensional form,

$$F_{sdm} = \frac{f_{sdm} h}{M_o C^2} = \sqrt{1-(V/C)^2} \qquad (4\text{-}9)$$

$$L_{sdm} = \frac{\lambda_{sdm} M_o C}{h} = \frac{V/C}{\sqrt{1-(V/C)^2}} \qquad (4\text{-}10)$$

Wave-particle Behavior of a Non-decaying Mass (NDM)

For a non-decaying mass m with a rest mass M_o and moving at velocity V, the equation (2-3) provides the following relationship,

$$m = \frac{M_o}{\sqrt{1-(V/C)^2}}$$

Substituting the above into equations (4-4) and (4-5), we obtain the following expressions for the frequency f_{ndm} and wavelength λ_{ndm} of the self-decaying mass,

$$f_{ndm} = \frac{M_o C^2}{h\sqrt{1-(V/C)^2}} \qquad (4-11)$$

$$\lambda_{ndm} = \frac{hV\sqrt{1-(V/C)^2}}{M_o C^2} \qquad (4-12)$$

These equations can also be expressed in the following non-dimensional form,

$$F_{ndm} = \frac{f_{ndm} h}{M_o C^2} = \frac{1}{\sqrt{1-(V/C)^2}} \qquad (4-13)$$

$$L_{ndm} = \frac{\lambda_{ndm} M_o C}{h} = \frac{V}{C}\sqrt{1-(V/C)^2} \qquad (4-14)$$

Comparison of de Broglie and GNM Based Wave-particle Models

Since the de Broglie model was based on classical or non-relativistic mechanics, the mass of a particle was considered to remain constant and not vary with the velocity. Similarly, the space and time were considered to be the fixed and invariant as in a Newtonian inertial frame. In GNM based wave-particle model, on the other hand, relativistic effects are built into equations (4-5) to (4-14).

Because of the relativistic effects, the wave-particle behavior predicted by the two models is different depending upon the mass and velocity of the particle and whether the particle is self-decaying or non-decaying. In order to understand these differences, we will investigate the behavior of some comparative quantities as follows. Equation (4-1) gives the deBroglie wavelength for a fixed mass M_0 as follows:

$$\lambda_{dbr} = \frac{h}{M_0 V} \qquad (4\text{-}15)$$

This can also be written in the following non-dimensional form:

$$L_{dbr} = \frac{\lambda_{dbr} M_0 C}{h} = \frac{1}{(V/C)} \qquad (4\text{-}16)$$

Now, de Broglie frequency can be calculated as follows,

$$f_{dbr} = \frac{V}{\lambda_{dbr}} = \frac{M_0 V^2}{h} \qquad (4\text{-}17)$$

or, in a non-dimensional form,

$$F_{dbr} = \frac{hf_{dbr}}{M_0 C^2} = \frac{V^2}{C^2} \qquad (4\text{-}18)$$

For a self-decaying mass, dividing equations (4-7) and (4-8) by equations (4-17) and (4-15) respectively, we obtain:

$$\frac{f_{sdm}}{f_{dbr}} = \frac{\sqrt{1-(V/C)^2}}{(V/C)^2} \qquad (4\text{-}19)$$

$$\frac{\lambda_{sdm}}{\lambda_{dbr}} = \frac{(V/C)^2}{\sqrt{1-(V/C)^2}} \qquad (4\text{-}20)$$

Similarly, for a non-decaying mass, dividing equations (4-11) and (4-12) by equations (4-17) and (4-15) respectively,

$$\frac{f_{ndm}}{f_{dbr}} = \frac{1}{(V/C)^2 \sqrt{1-(V/C)^2}} \qquad (4\text{-}21)$$

and,

$$\frac{\lambda_{ndm}}{\lambda_{dbr}} = (V/C)^2 \sqrt{1-(V/C)^2} \qquad (4\text{-}22)$$

Figure 4-1 shows variation of non-dimensional frequencies calculated by GNM equation (4-9) for a self-decaying mass, equation (4-13) for a non-decaying mass and de Broglie equation (4-18) as a function of velocity ratio V/C. At V=0, the non-dimensional frequency calculated by de Broglie model is zero, while GNM calculated non-dimensional frequencies are equal to 1. As V increases to equal C, the de Broglie frequency increases to reach a maximum of 1, while the frequency for the self-decaying mass decrease to zero because of a total mass dilation to energy and the frequency for the non-decaying mass increases indefinitely because of the addition of external energy required to accelerate its velocity to C.

Figure 4-2 shows variation of non-dimensional wavelengths calculated by GNM equation (4-10) for a self-decaying mass, equation (4-14) for a non-decaying mass and de Broglie equation (4-16) as a function of velocity ratio V/C. At V=0, the non-dimensional wavelength calculated by de Broglie model is infinite, while GNM calculated non-dimensional wavelengths are equal to 0. As V increases to equal C, the de Broglie wavelength decreases to reach a minimum value of 1, while the wavelength for the self-decaying mass increases indefinitely because of a total mass dilation to energy and the wavelength for the non-decaying mass decreases back to zero because of infinite increase in its mass due to the addition of external energy required to accelerate its velocity to C.

It should be noted that because of the enhanced relativistic effects, the predictions of GNM and de Broglie models are expected to give very different results as V approaches C. At small velocities (V~0), however, de Broglie model predicts infinite wavelength even for a large mass, which is counter to the physical experience. Large objects when stationary (V=0) or moving at very slow speeds do not act or appear as a wave. Hence, de Broglie model does not appear to represent physical reality at low values of V. This discrepancy is understandable since this model was originally developed for quantum particles of small masses, such as photons, moving at large velocities close to C. For example at V=0.8C the predictions of de Broglie model and GNM model for a self-decaying mass are fairly close as shown in Figures 4-1 and 4-2. Figure 4-3 shows the predicted wavelengths of an electron with a rest mass of 10^{-30} kilogram using different models. As discussed earlier, a good agreement is seen between GNM self-decaying mass model and de Broglie model for V in the range of 0.7C and 0.9C. The maximum wavelength predicted by GNM non-decaying mass model for an electron is approximately 10^{-12} meters. Figure 4-4 shows calculated wavelengths for a proton with a rest mass of 1.6×10^{-27} kilogram.

The Hidden Factor

The ratio of wavelengths predicted by GNM to de Broglie models for a self-decaying mass, equation (4-20) and non-decaying mass, equation (4-22) are shown in Figure 4-5. At V=0, the ratio is zero because GNM model wavelength is zero. As V increases to C, the ratio approaches zero for the non-decaying mass as its GNM wavelength goes to zero and infinity for the self-decaying mass since its GNM wavelength becomes infinite. Again, a good agreement is seen between the deBroglie model and GNM model for self-decaying mass at V/C of approximately equal to 0.8.

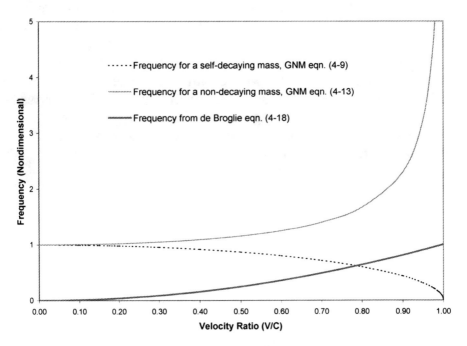

Figure 4-1: Comparison of non-dimensional frequencies predicted by GNM and de Broglie models.

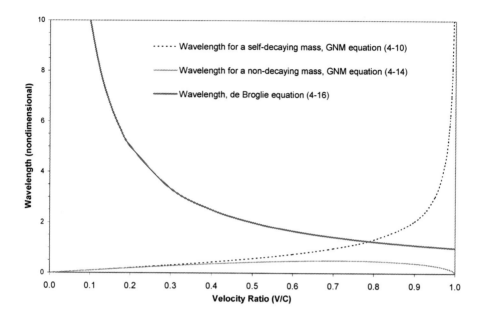

Figure 4-2: Comparison of non-dimensional wavelengths predicted by GNM and de Broglie models.

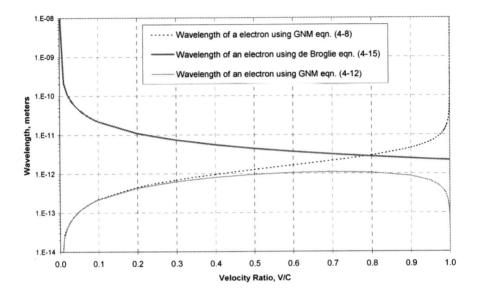

Figure 4-3: Comparison of wavelengths predicted by GNM and de Broglie models for an electron.

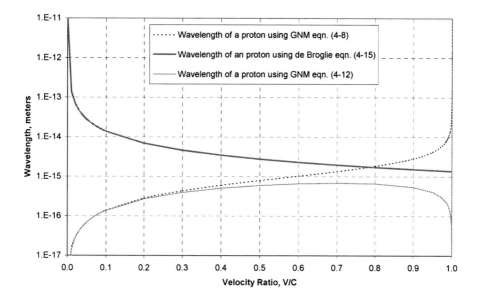

Figure 4-4: Comparison of wavelengths predicted by GNM and de Broglie models for a proton.

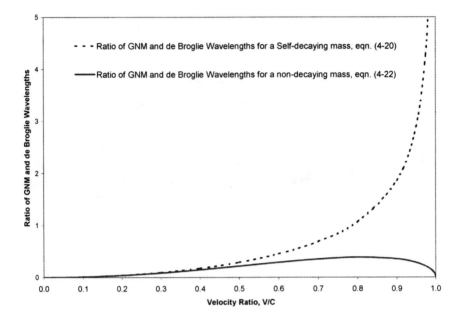

Figure 4-5: Ratio of wavelengths predicted by GNM and de Broglie models.

Visibility of an Entity as a Rigid Body

The wave-particle duality is expected to have an impact on the physical behavior, specifically its visibility as a rigid body with well-defined boundaries that are spatially separated from the environment. Ordinary objects in the real world are visible and perceived to have spatial boundaries. Their space-time location can be specified in three-dimensional space and time coordinates in a Newtonian inertial frame of reference. The mass of ordinary worldly objects is so large that their equivalent wavelength given by equation (4-12) for a non-decaying mass is much smaller than their physical size. The effect of mass and size on the matter wavelength of an object is shown in Figure 4-6.

As an example, a stationary ($V=10^{-9}$ meters/sec) drop of water weighing one gram has a diameter of about 1.24 centimeter and a wavelength of approximately 2.2×10^{-46} centimeter using equation (4-12). Because of such a small wavelength the wavy nature of the drop is insignificant and the drop is perceived to have well defined boundaries identifiable and distinguishable from the environment of space. But a much smaller drop that weighs only 10^{-50} kilograms has a diameter of 2.67×10^{-16} centimeters (assuming a constant density of 1 kilogram per liter) and a calculated wavelength of 22 centimeters. For the purpose of simplifying the discussion of the dualistic wave-particle behavior of real world objects, we have assumed that the atomic structure and the physical properties such as the density of the drop are maintained at small sizes. For extremely small drops, the wave-like behavior completely dominates the particle-like behavior and the drop will not be visible or perceived as a rigid body with well-defined and separated boundaries. Also, at any given time, it is not possible to prescribe a specific point-like spatial location for such a small drop in an inertial frame of reference since the drop could be located anywhere in a region of space equal to its wavelength which is several orders of magnitude larger than its diameter.

Figure 4-7 shows the variation of water drop sizes and corresponding wavelengths calculated from equation (4-12) for a wide range of rest mass values for stationary ($V=10^{-9}$ meters/sec) drops. It should be noticed from Figure 4-7 that for a drop of a rest mass equal to 5×10^{-38} kilogram, both the drop size and its wavelength are approximately equal to 4.5×10^{-12} centimeter. The mass of such a drop could be defined as the critical mass determining the wave-like or particle-like behavior of the drop. A drop with a mass less than the critical mass will behave like a wave and a drop with a higher mass will behave like a particle or a rigid body.

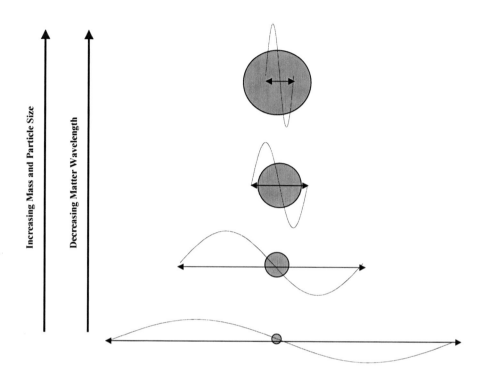

Figure 4-6: GNM predicted wavelength versus particle mass and size.

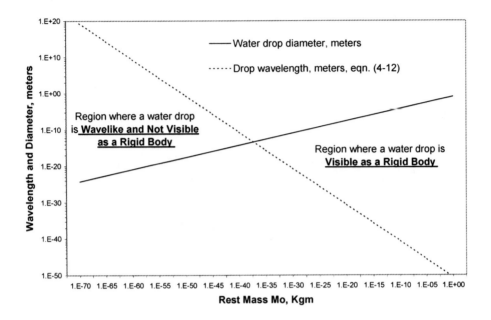

Figure 4-7: Wavelength and size versus mass of a drop of water at low velocity.

The wavelength of visible light is approximately 5×10^{-7} meters. We can assume that a body that has a size larger than the wavelength of the visible light will be visible to the human eye with an optical microscope of a proper magnification capability. Using equations (4-8) and (4-12), we can express the rest mass for a given wavelength using GNM for a self-decaying and non-decaying mass as follows:

$$M_{0sdm} = \frac{hV}{\lambda_{sdm} C^2 \sqrt{1-(V/C)^2}} \quad (4\text{-}23)$$

$$M_{0ndm} = \frac{hV\sqrt{1-(V/C)^2}}{\lambda_{ndm} C^2} \quad (4\text{-}24)$$

Similarly, using equation (4-15) and de Broglie model,

$$M_{0dbr} = \frac{h}{\lambda_{dbr} V} \quad (4\text{-}25)$$

Figure 4-8 shows the rest masses calculated by GNM equations (4-23) and (4-24) for a self-decaying and non-decaying mass respectively, and equation (4-25) using de Broglie model for a specified wavelength of 5×10^{-7} meters. At small velocities, the rest mass calculated by GNM is very small while the de Broglie model calculates a very large rest mass. As velocity increases, the rest mass calculated by GNM increases indefinitely for a self-decaying mass but reaches a maximum and then decreases to zero for a non-decaying mass as V approaches C. For de Broglie model, as V increases the calculated rest mass decreases to a constant value of approximately 5×10^{-35} kilogram at V=C.

It is interesting to note that a self-decaying mass, no matter how large its initial rest mass may be, will completely convert to an energy wave as V approaches C. A non-decaying mass, on the other hand, will behave like a particle of infinite mass as V approaches C. It will require an infinite amount of external energy input to accelerate a non-decaying mass to attain a speed C and hence, this state is physically impossible to achieve.

Figure 4-9 shows comparison of the rest masses calculated by GNM equations (4-23) and (4-24) for a self-decaying and non-decaying mass respectively for two different specified wavelengths of 5×10^{-7} meters and 1×10^{-10} meters. As expected, the rest mass calculated for the smaller wavelength is larger since they are inversely proportional to each other.

Again, there is a reasonably good agreement between GNM and de Broglie models when V is close to 0.8C. However, it should be noted that the results of the de Broglie model at V=0, do not match the observed reality. The model predicts that a very large mass at V=0 will have a large wavelength and hence will act like a wave. The large stationary masses we observe in our common experience in the world act like separated bodies with well-defined and spatially identifiable boundaries. In fact, the bigger and heavier the body is, the less wavelike it acts. This is counter to the predictions of the de Broglie model. This aspect of the de Broglie model is also counter-intuitive to the Heisenberg Uncertainty and quantum mechanics theories, as we will discuss later in the following chapters.

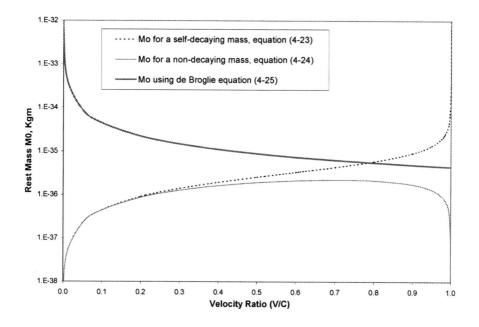

Figure 4-8: Rest mass values predicted by GNM and deBroglie models for a wavelength of 5×10^{-7} meters.

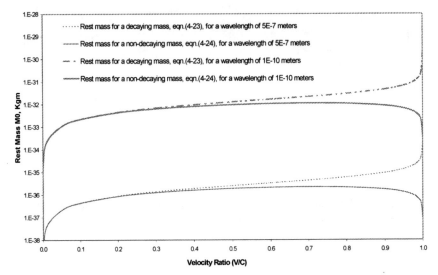

Figure 4-9: Rest masses predicted by GNM for wavelengths of 5×10^{-7} & 1×10^{-10} meters.

The maximum value of the rest mass for a given wavelength predicted by GNM for the non-decaying mass can be derived from equation (4-24), which is rewritten in the following non-dimensional form:

$$\frac{M_{0ndm} \lambda_{ndm} C}{h} = \frac{V}{C}\sqrt{1-(V/C)^2} \qquad (4\text{-}26)$$

In order to find the maximum value of the left-hand side in the above equation, we set the first derivative of the above equation with respect to V/C equal to zero. This yields the following:

$$\left(\frac{M_{0ndm} \lambda_{ndm} C}{h}\right)_{max} = 0.486$$

or,

$$(M_{0ndm})_{max} = 0.486 \frac{h}{\lambda_{ndm} C} \qquad (4\text{-}27)$$

This maximum occurs at V/C= 0.618. Figure 4-10 shows the maximum value of a non-decaying mass, given by equation (4-27) for different values of its wavelength.

Effect of Rest Mass on Frequency and Wavelength

The effect of varying the rest mass on frequency and wavelength for a self-decaying mass given by equations (4-7) and (4-8) respectively and for a non-decaying mass given by equation (4-11) and (4-12) respectively are shown in Figures 4-11 for two different masses of 1×10^{-35} kilogram and 1×10^{-45} kilogram. The two upper curves in Figure 4-11 depict frequencies for heavier of the two masses and the lower curves depict frequencies for the lighter mass. Since the frequency is proportional to the mass-energy of the body, the higher mass has higher frequency. At V=0, the self-decaying mass and the non-decaying mass have equal frequencies. As V increases, frequency predicted by GNM remains fairly constant until V approaches C. At V=C, the frequency for the non-decaying mass increases indefinitely due to an infinite increase in mass while the frequency for the self-decaying mass dilates to zero via converting to wave energy. Figure 4-12 shows wavelengths corresponding to frequencies shown in Figure 4-11 for the two masses. The effect of the rest mass on wavelength is opposite to that on frequency, since the wavelength is inversely proportional to the rest mass.

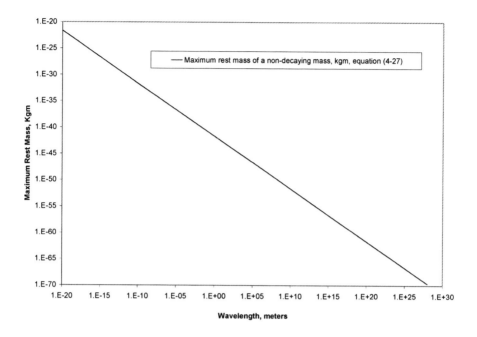

Figure 4-10: Maximum rest mass values of a non-decaying mass predicted by GNM model, equation (4-27).

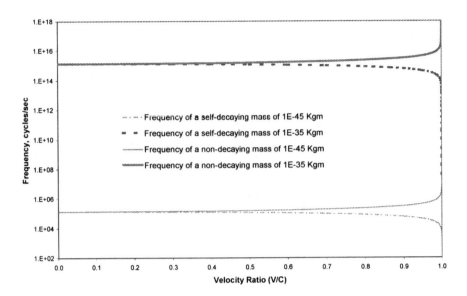

Figure 4-11: Frequencies predicted by GNM for self-decaying and non-decaying masses of 1×10^{-35} kilogram and 1×10^{-45} kilogram.

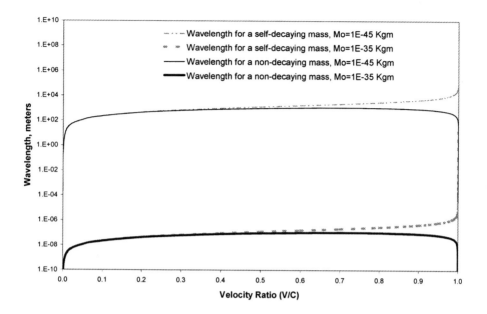

Figure 4-12: Wavelengths predicted by GNM for self-decaying and non-decaying masses of 1×10^{-35} kilogram and 1×10^{-45} kilogram.

Avtar Singh

Wave-particle Behavior of a Photon During Absorption and Emission

When does a mass behave as a particle and under what conditions it behaves like a wave? It is generally believed that the particle behavior of a photon is observed when it hits a detection surface or screen. The concept of a Photon as a particle was proposed by Einstein to explain the observed photoelectric effect wherein the energy of absorbed light quanta can cause an emission of electrons from a surface. The wavelike behavior is observed during its uninterrupted travel or flight from the source to a detection screen or during its motion through a slit or around a solid edge. In other words, when the motion of the photon is stopped or interrupted completely, it acts as a particle. On the other hand, if the motion is maintained or allowed to continue, even if partially, the wave behavior is prevalent.

It is generally assumed that a photon has zero mass and velocity equal to the speed of light C. It is also assumed to have energy equal to the product of Planck's constant h and frequency f of the light wave. These assumptions are not physically consistent as follows. The mass of a photon is complimentary to its energy and hence, can not be zero if the energy of the photon is non-zero. For a specific energy of a photon, the equivalent mass of the photon is given by the ESTR relation- $E=mC^2$. For a photon with a rest mass of M_0, this leads to the following:

$$E = hf = M_0 C^2$$

or,

$$f = \frac{M_0 C^2}{h} \qquad (4\text{-}28)$$

The wavelength of the photon is given by:

$$\lambda = \frac{C}{f} = \frac{h}{M_0 C} \qquad (4\text{-}29)$$

Figure 4-13 shows the frequencies and wavelengths predicted by equations (4-28) and (4-29) as a function of varying rest mass of a photon. As the rest mass decreases, the frequency decreases and the wavelength increases. This does not support the assumption that a photon has zero rest mass. Another evidence of the non-zero mass of the photon is the observed bending of light by gravity force from a large body. Einstein's general theory of relativity also predicts such bending of light by gravity.

Another inconsistency in the assumptions above is that the velocity of a photon at the moment of its emission or absorption at a stationary surface must be zero and not C. Following its emission from the stationary surface, a photon may accelerate away to the speed of light when it's motion is not constrained by the stationary surface of emission. Two questions then arise as to how this acceleration occurs. The first question is- 'How does the photon get accelerated to achieve high velocities?' Since, there are no external forces acting on the photon, such a motion has to be internally induced. A self-decaying mass proposed by GNM does have such a characteristic of spontaneous motion and would provide a possible explanation of the observed photon behavior. The second question is- 'If the mass of a photon is non-zero due to its finite non-zero energy, then according to ESTR it would take an infinite amount of energy to accelerate it to achieve the speed of light C. Where does this infinite energy come from?' Again, since there is no external source of energy to move the photon, the energy for its motion must be internally induced. And this supports the existence of the spontaneous self-decaying nature of the photon mass as conceptualized in GNM.

Figure 4-14 provides a physical picture of the photon behavior as a self-decaying mass during the processes of its emission or absorption at a stationary surface. At the surface, its velocity is zero to match the stationary boundary condition and its particle rest mass is equal to the equivalent quantum energy of the photon. During its emission, the mass of a photon spontaneously decays to provide kinetic energy for its motion away from the surface. The velocity of the photon accelerates as the mass converts to kinetic energy until it decays completely to attain the speed of light C. The self-decaying characteristic of the photon mass negates the need for any external force or source of energy to accelerate the photon to the speed of light and provides a physically consistent explanation of its spontaneous wave-particle behavior. During the absorption process of an incoming photon at a stationary surface, the above sequence of events is reversed wherein the velocity of the photon decreases from C to zero and the mass increases from zero to the rest mass.

Frequencies and wavelengths of a self-decaying photon with a rest mass of M_0, can be calculated using equations (4-7) and (4-8) for varying values of velocity. Figure 4-15 and 4-16 show comparisons of photon frequencies predicted by equations (4-7) and (4-28) for two different photon masses of 1×10^{-45} kilogram and 1×10^{-35} kilogram respectively. At small velocities, the two models predict an equal and constant frequency. As velocity increases and approaches to C, the frequency predicted by the self-decaying model equation (4-7) decreases to zero while the frequency predicted by equation (4-28) remains constant. Since the frequency is proportional to the rest mass, the frequency for the heavier rest mass is proportionally higher than the lighter one. Figure 4-17 shows the ratio of frequencies predicted by GNM equation (4-7) for a self-decaying photon and frequency given by equation (4-28).

Figure 4-18 and 4-19 show comparisons of photon wavelengths predicted by equations (4-8) and (4-29) for two different photon masses of 1×10^{-45} kilogram and 1×10^{-35} kilogram respectively. At zero velocity, the wavelength predicted by GNM equation (4-8) is zero and as V increases to C the wavelength increases to infinity. The wavelength predicted by equation (4-29) remains constant since it does not depend upon V. Since the wavelength is inversely proportional to the rest mass, the wavelength for the heavier rest mass is proportionally lower than the lighter one. Figure 4-20 shows the ratio of wavelengths predicted by GNM for a self-decaying photon equation (4-8) and photon model given by equation (4-29). Good agreement between the two models exists when V is in the vicinity of 0.7C.

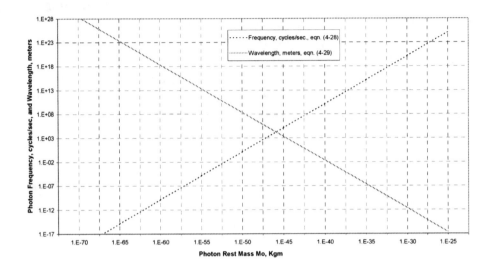

Figure 4-13: Wavelength and frequency of a photon for different rest masses.

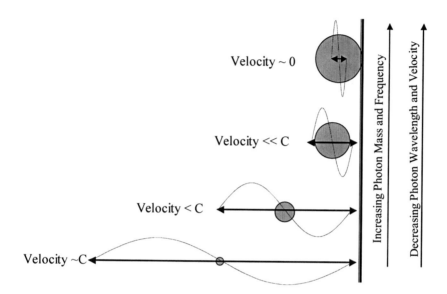

Figure 4-14: Wave-particle behavior of a photon during absorption and emission.

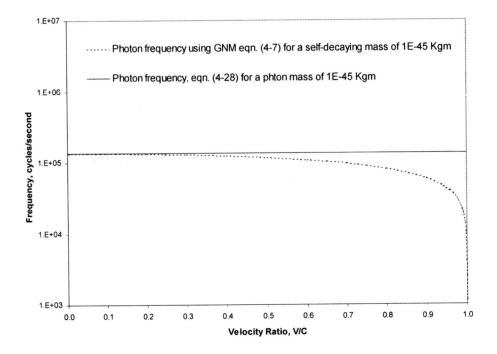

Figure 4-15: Frequencies of a photon predicted by GNM for a self-decaying mass of 1x10⁻⁴⁵ kilogram.

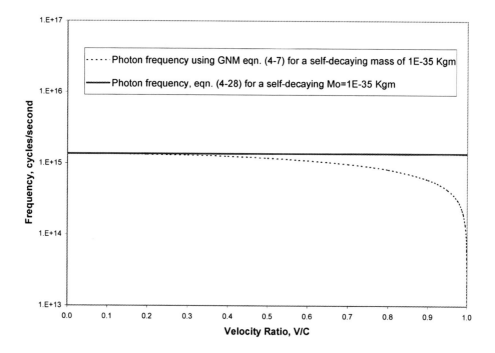

Figure 4-16: Frequencies of a photon predicted by GNM for a self-decaying mass of 1x10^{-35} kilogram.

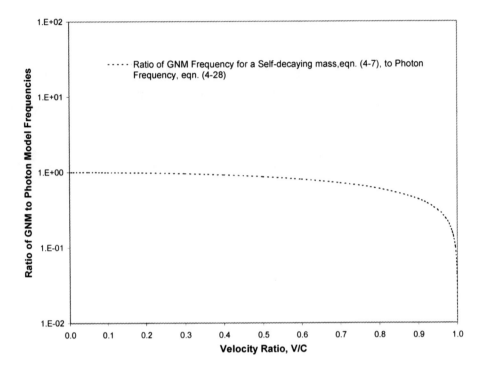

Figure 4-17: Ratio of frequencies of a photon predicted by GNM for a self-decaying mass and classical photon model.

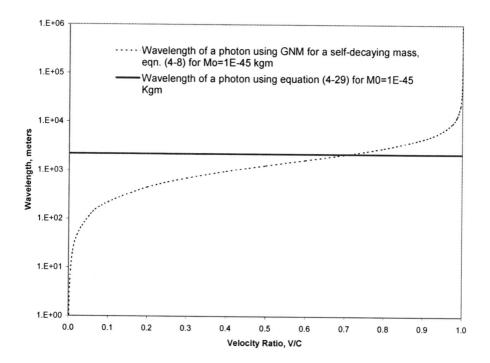

Figure 4-18: Wavelengths of a photon predicted by GNM for a self-decaying mass of 1×10^{-45} kilogram.

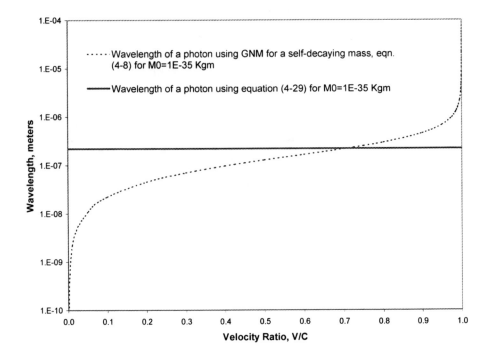

Figure 4-19: Wavelengths of a photon predicted by GNM for a self-decaying mass of 1×10^{-35} kilogram.

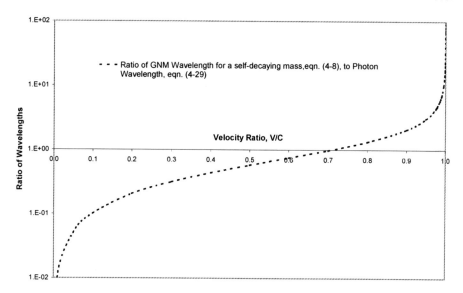

Figure 4-20: Ratio of wavelengths of a photon predicted by GNM for a self-decaying mass and classical photon model.

Avtar Singh

Wave-particle Duality and Non-locality

Since the wave-particle behavior and wavelength predicted by GNM for a given body are both dependent upon its mass and velocity, a strong correlation is expected to exist between the wavelength and non-local behavior. We will now discuss the relationship between the wavelength and non-locality using GNM equations for wavelength in this chapter for both self-decaying and non-decaying masses.

Wavelength of a self-decaying body with a rest mass M_o is given by equation (4-8),

$$\lambda_{sdm} = \frac{hV}{M_o C^2 \sqrt{1-(V/C)^2}} \quad (4\text{-}8)$$

As a self-decaying mass dilates to zero when its velocity V approaches C, its wavelength predicted by equation above increases to an infinite value, thus physically expanding over the entire universe. This explains the observed non-locality of a photon moving in an empty space. Because of the infinite wavelength, the photon exhibits a complete coherence in its inherent properties measured over large distances in the stationary frame of reference of the quantum entanglement experiments performed by scientists.

In chapter 2, we demonstrated that the non-locality results from conservation of space-time when the speed V of an entity or its frame of reference equals C. In chapter 3, we provided a more generic basis for non-locality, which is the conservation of mass-energy in the universe. In this chapter, we provided another physical justification for non-locality based on an infinite wavelength attained by a self-decaying mass as its mass dilates to zero at V equals C.

It is argued and implied in ESTR that a real signal that carries information in the form of an energy wave with a finite non-zero frequency and wavelength can never travel faster than C. This is consistent with wavelength predicted by the Gravity Nullification Model equation (4-8) as follows. The information in a signal is stored in the form of energy waves of specific frequency and wavelength as seen by an observer in the stationary frame of reference. Since the signal maintains a finite energy or non-zero rest mass that constitutes the specific stored information during its transmission through empty space, its speed V remains slightly less than (however very close to the value of) C and the corresponding wavelength predicted by equation (4-8) remains relatively small depending upon the total energy and equivalent value of its rest mass. This preserves the locality aspects of the signal since its mass is larger than zero and velocity V is slightly smaller than C. A photon of light, on the other hand that is not burdened with a signal of a non-zero rest mass, can decay its mass to zero as it travels through the empty space. Hence, non-locality and coherence is observed in the behavior of a photon over the entire universe.

For a non-decaying mass, wavelength is given by equation (4-12):

$$\lambda_{ndm} = \frac{hV\sqrt{1-(V/C)^2}}{M_o C^2} \quad (4\text{-}12)$$

As velocity increases and approaches C, the wavelength of a non-decaying mass decreases as shown in Figure 4-12. Hence, a non-decaying mass will always exhibit a local rather than non-local behavior.

Chapter 5

A Universe Model based on the Gravity Nullification Model

In Chapter 3, we derived GNM based on the holistic mass-energy conservation and spontaneous motion of a self-decaying mass. In Chapter 4, we used GNM to explain the observed and spontaneous wave-particle behavior of quantum particles in the universe. In this Chapter we will develop a GNM based model of the universe to explain the observed behavior of the visible as well as invisible universe i.e. Dark Matter. We will first review the widely accepted universe model, the Big Bang Model, based on Newtonian gravity and general relativity theories. Then we will use GNM to explain the remaining gaps in the physical understanding of singularities in initial conditions (the Big Bang), evolution or history of the universe, multiple universes, cosmological constant, acceleration of the universe expansion, it's future and philosophical aspects related to the anthropic principle etc. We will demonstrate how GNM can provide consistent and mechanistic insights into the majority of the open questions that still persist in spite of the highly successful general relativity and quantum mechanics theories.

The Big Bang Model (BBM)

The 'Big Bang' model [4] of the universe is based on the observed expansion of the universe. According to the Hubble's law, each distant galaxy is receding from us with a velocity proportional to its distance. If the universe has been expanding as observed by Hubble, it must have been smaller in the past, and also denser and hotter. Hubble's law gives the velocity of recession V at a distance R as follows:

$$V = HR \tag{5-1}$$

H in the above equation represents the expansion rate of the universe and is known as the Hubble's constant (H is called a constant since it does not vary with R but according to the Newtonian model of the universe its value does vary with time). In its current form, the Big Bang model is based on mathematical solution of the equations of general relativity, originally obtained by Friedman in the 1920s. This model is based on the general cosmological principle that on large scales, the universe is homogeneous (uniform density of matter), looks the same in every direction (isotropic) with each particle moving according to equation (5-1). The total energy E_T of a particle of mass m at a distance R is then given by,

$$E_T = \frac{1}{2}mV^2 - \frac{GMm}{R} \tag{5-2}$$

wherein, G is the Newton's Gravitational Constant and mass M is given in terms of the uniform mass density ρ as follows,

$$M = \frac{4\pi}{3}R^3\rho \tag{5-3}$$

Equation (5-3) can be rearranged in the following non-dimensional form,

$$\frac{2E_T}{mV^2} + \frac{2GM}{RV^2} = 1 \tag{5-4}$$

Substituting equations (5-1) and (5-3) into equation (5-4) gives,

$$\frac{2E_T}{mH^2R^2} + \frac{8\pi G\rho}{3H^2} = 1 \tag{5-5}$$

or,

$$\Omega_k + \Omega_m = 1 \tag{5-6}$$

wherein, $\Omega_k = \dfrac{2E_T}{mH^2R^2}$ represents non-dimensional curvature of the universe and $\Omega_m = \dfrac{8\pi G\rho}{3H^2}$ represents the critical mass density ratio.

Since the gravity slows the expansion of the universe, its expansion rate has not been constant. If matter is dense enough, $\Omega_m > 1$, the gravity force of attraction will eventually halt the expansion and cause the universe to collapse in a Big Crunch. This is called a "closed' universe, like a sphere. If mass density is too small, $\Omega_m < 1$, the kinetic energy will take over the force of attraction causing the universe to expand forever. This is called an open universe, like a hyperboloid or saddle. When $\Omega_m = 1$, the mass density is equal to the critical density given by the following equation:

$$\rho_{crit} = \dfrac{3H^2}{8\pi G} \qquad (5\text{-}6)$$

The critical density universe will expand without accelerating and this is known as the flat universe wherein Euclid's theorems apply.

The above models apply to a universe with empty space containing no inherent energy. Einstein proposed a 'Cosmological Constant' denoted by Λ, which represented a contribution to the density of the universe from vacuum energy. If Λ is greater than zero, then a spatially flat universe with low matter density can exist due to the contribution from the vacuum energy. An accelerating universe can exist only when Λ is greater than zero. For a non-zero Λ, Einstein proposed the following modification to equation (5-2):

$$E_T = \dfrac{1}{2}mV^2 - \dfrac{GMm}{R} - \dfrac{1}{6}\Lambda mC^2R^2 \qquad (5\text{-}7)$$

Equations (5-5) and (5-6) can be rewritten to include Λ as follows:

$$\frac{2E_T}{mH^2R^2} + \frac{8\pi G\rho}{3H^2} + \frac{\Lambda C^2}{3H^2} = 1 \qquad (5\text{-}8)$$

$$\Omega_k + \Omega_m + \Omega_\Lambda = 1 \qquad (5\text{-}9)$$

wherein, $\Omega_\Lambda = \dfrac{\Lambda C^2}{3H^2}$ represents the vacuum energy density ratio.

Shortfalls of the Big Bang Model (BBM)

The Big Bang model has been successful in explaining a number of features of the observed universe in an irksome marginal way. The features explained include the cosmic microwave background radiation and the abundances of the lighter elements through the process of nucleosynthesis. It also provides an explanation of how the large structures, such as galaxies, form in a very early homogeneous universe. Some big questions [10, 11, 14, 16] still remain unanswered and inconsistencies persist that could invalidate the model in several ways discussed below:

1. Singularity in the initial conditions

The numerical values of the Big Bang model parameters at the beginning of time (t=0) can not be specified because the physical laws break down at the origin of time due to a physical singularity. This is a serious shortcoming of BBM, since physical singularities should be avoided in all physically meaningful theories. Roger Penrose [33] states-

"All these models (Friedman models of the universe) have this awkward feature, known as the Big Bang, where everything goes wrong, right at the beginning. The universe is infinitely dense, infinitely hot and so on – something has gone badly wrong with the theory."

"What is the probability that, purely by chance, the universe had an initial singularity looking even remotely as it does? The probability is less than one part in $10^{10^{123}}$.....If I were to put one zero on each elementary particle in the universe, I still could not write the number down in full. It is stupendous number."

Most cosmologists, however, believe that BBM provides the most convincing understanding of the observed large-scale features of the universe. The predicted results, however, come out right only if the initial conditions are arranged rather incredibly carefully. The initial conditions must be inferred from extremely large extrapolations of the observed physical behavior of the universe. It is not certain that the physical phenomena existing at the beginning of time could ever be comprehensively described, which makes these initial conditions highly speculative. The history and evolution of the universe up to modern times strongly depends upon the initial conditions, and since they are not known the whole evolution of the universe essentially becomes empirical without any help from the Big Bang model.

Another key question is what set the universe in motion? The cosmological expansion is the generic feature of the Einstein's Theory of general relativity and the Newtonian approximations depicted by the equations (5-1) thru (5-9) above. The motion is thus enforced by prescription of governing equations without a solid explanation of the mechanistic physics behind it.

2. Uncertainty in the historical and current values of parameters

The Hidden Factor

The BBM accounts three significant aspects of the current universe: the expansion according to the Hubble's law given by equation (5-1), the existence of 2.7 ^{0} K micro wave background radiation in all directions of the sky, and the abundance of light atomic nuclei (helium and deuterium, which are postulated to be created in the very early universe. The uncertainty induced by the scatter and varied interpretation of the available data to quantify the parameters (Ω_k, Ω_m and Ω_Λ) of the model remains very large reducing the confidence in the predictions of the model. In the current solutions of the Big Bang model, for reasons of simplicity it is assumed that these parameters have remained constant during the evolution of the universe from the occurrence of the Big Bang to the current time period. However, there is no solid evidence available from the observations for supporting or refuting this leap of faith. Even if these parameters were known to vary with time over the evolution of the universe, it would be impossible to define or quantify that variation with the available observations today.

3. Unavoidable but Incredible Inflation Scenario to make Big Bang model work:

Even though the Big Bang model explained several aspects of the observed universe, a large coincidence and fine-tuning is still required to explain three major problems [11]. These are described below:

The Horizon Problem

Because of the limitations of the speed of light, there would be regions in space, which will remain invisible to an observer because the light from the origin has not arrived as yet in those regions. This creates the horizon problem since such a restriction does not seem to apply to the microwave background radiation, which is observed to be incredibly homogeneous and uniform in temperatures and density without any indication of inaccessible regions in the universe. How such homogeneity exists in the universe? A limiting distance determined by how far the light has traveled must exist beyond which nothing can be seen. The horizon effect prevents any mechanistic and physical reason for such a remarkable communication and homogeneity in the properties of the universe so far as the galaxies and the cosmic background radiation are concerned. The incredible homogeneity observed in the background radiation to an accuracy of 3 parts in 100,000 creates an astronomical mystery that fails an appealing explanation. This implies that BBM needs to be supplemented by another physical model that would allow enhanced communication (at speeds which are several times faster than the speed of light) among various early and later regions of the universe.

The Flatness Problem

Observations of the current universe show that it is flat without any significant positive or negative curvature. Since both the mass as well as the vacuum energy of the Cosmological Constant contribute to the curvature of the universe, a flat universe implies that the sum of Ω_m and Ω_Λ is close to 1. The flatness problem arises from the incredible fact that this sum is close to 1 when it could theoretically be any number different from 1. The solution of the Big Bang model equations would predict a stable flat universe only if it was flat initially at t=0. This implies that the initial condition of flatness needs to be imposed on the model to match the observed behavior of the universe.

Another implication of the flat universe is that the total energy of the universe, E_T, must remain equal to zero at all times during its evolution. This also leads to the assumption that there must exist equal amounts of matter and anti-matter in the universe to exactly annihilate each other to make the net energy equal to zero. This is not at all consistent with the observations and human experience of the universe, which appears to be dominated by matter and filled with visible stars and galaxies without any direct evidence of the presence of anti-matter.

The Cosmological Constant Problem

Einstein introduced the Cosmological Constant in his equations of general relativity to ward off expansion and to obtain a widely favored static universe. The value that Einstein assigned to the constant was the one that provided a static universe without any expansion or contraction. The current understanding of the Cosmological Constant represents existence of a negative pressure caused by the energy of the vacuum. Its sole function is to push the space radially outwards causing it to expand. As the space expands it accumulates more and more of the vacuum energy, which eventually overcomes gravity and results in an accelerating universe. It contributes to the mass or energy density and hence to the gravity term in the model.

Since the original intent for the Cosmological Constant was to achieve a static universe, when observations by Hubble discovered an expanding universe, Einstein called it "the Biggest Blunder". However, the observations of the universe today strongly support a non-zero Cosmological Constant along with a flat universe. As the universe expands, the Gravity force diminishes due to larger distances and the Cosmological Constant dominates the accelerating expansion. Quantum mechanics predicts that the so-called vacuum of the empty space is filled with elementary particles that pop in and out of existence too quickly to avoid a physical detection. So far no credible physical theory has been advanced that could predict even the right order of magnitude of the aggregate vacuum energy of these unstable or self-decaying particles. This vacuum energy could represent the expansion energy corresponding to the Cosmological Constant. Particle-physics theories put forward possible partial explanation for existence of vacuum energy and a non-zero Cosmological Constant, but the values predicted by these are about 120 orders of magnitude greater than the recent supernova observations. If the vacuum energy of such magnitude were existent, an acceleration of almost infinite magnitude would rip apart atoms, stars and galaxies. Clearly, the current understanding of the vacuum energy is incorrect. The inaccuracies get even more magnified when the existing theories are extrapolated to the beginning time of the universe. The current values of the Ω_m and Ω_Λ are close to 1. At much earlier times, the universe volume was 100 orders of magnitude smaller, and hence Ω_m would have been 100 orders of magnitude larger than Ω_Λ, if it remained constant as assumed in BBM. How such a huge disparity would lead to an evolution of the universe as it exists today would be a matter of utter coincidence, or the so-called fine-tuning, which is impossible to be accepted by rational scientific arguments. Since BBM assumes that Ω_m remains constant and close to 1, it implies that the actual mass density of the universe stays close to the critical mass density given by equation (5-6). This requires the mass of the universe to increase in direct proportionality to its volume as the universe expands. However, since the net mass-energy of the universe is assumed to remain zero, the amount of anti-matter has to increase the same as the visible matter. None of these predictions of BBM are consistent, or at least verifiable, with the actual observations of the universe.

There are other unanswered questions regarding the Cosmological Constant. One question is whether the Cosmological Constant is really a universal constant that does not change with the time evolution of the universe. Recently, some scientists [26] have proposed a variable (Quintessence) Cosmological Constant to account for the behavior of the observed data.

Another puzzle, also known as the "cosmic coincidence" problem that haunts scientists is why the Cosmological Constant comes to dominate at about the epoch of galaxy and star formation and not earlier. This is counter-intuitive to the notion that prime mover for galaxy and star formation is the pull of gravity rather than the expansion induced by ant-gravity signified by the Cosmological Constant. If the Dark Energy is made up of the vacuum energy, it is impossible to account for such a coincidence. Is there a deeper connection between the evolution of thinking beings and the onset of the rampant acceleration of the universe? Recently [3], the Anthropic Principle, which has a low approval rating among physicists, has been shown to be the only approach that can naturally resolve the problems related to the Cosmological Constant. The Anthropic Principle is based on the assumption that among all the possible multitude of universes, the intelligent beings capable of contemplating could exist only in the current "best of all worlds". Thus the puzzle is locked into a one way and dead-end street constituting a big credibility gap in accepting BBM.

Inflation - a Savior (sort of) of the Big Bang Model

Inflation is a so-called "superluminal expansion" of the universe that is presumed to occur during the very early universe when it was about 10^{-35} seconds old. The vacuum energy of the universe, which later condenses into ordinary matter, fuels inflation. The Inflation solves the horizon and the flatness problems by stretching the universe to a phenomenal size almost instantaneously, so the expanded distance now determines the horizon size that is many orders of magnitude larger than the distance traveled by light at its normal speed. Because of this almost infinite stretching, the deviation from flatness also becomes negligible. Inflation thus essentially presumes that the contents of the universe move at speeds much greater than the speed of light violating the Einstein's theory of specific relativity. *The price to pay for this incredible solution is thus to sacrifice the maximum speed limit in the universe.*

Inflation solves the horizon and the flatness problems, however it does not solve any of the Cosmological Constant problems mentioned above. In fact, the inflation relies on the Cosmological Constant or the vacuum energy to fuel the rapid expansion, without providing a mathematical or mechanistic model to evaluate its magnitude. *Inflation thus intensifies the need to find a credible technical basis to calculate the Cosmological Constant, especially during the post-inflation regime.*

Roger Penrose [33] states the following regarding the theory of Inflation -

> "But, as it stands, the argument does not do what it is supposed to do – what you would expect in this initial state, if it were randomly chosen, would be a horrendous mess and, if you expand that mess by this huge factor (10^{60}), it still remains a complete mess. In fact, it looks worse and worse the more it expands."

4. Singularity in the observed accelerated expansion conditions

The singularity in BBM often not realized or highlighted by the scientific community is the infinitely increasing mass as the universe expansion accelerates to velocities close to the speed of light. As predicted by the theory of relativity, the mass increases with velocity as follows:

$$M = \frac{M_o}{\sqrt{1-(V/C)^2}} \qquad (2\text{-}3)$$

Most commonly used velocity distribution in existing Big Bang models is the Hubble's law, which describes the velocity of recession V at a distance R as follows:

$$V = HR \qquad (5\text{-}1)$$

At large values of R, V increases without limit and as V approaches C, the mass calculated by equation (2-3) reaches an infinite value causing a singularity. When V calculated by equation (5-1) becomes larger than C at even larger R, it violates the specific theory of relativity.

The above shortcomings and pitfalls of the Big Bang model have lured several investigators to look into alternative models. Now, we will focus on how the Gravity Nullification Model

(GNM) described in the previous chapters can be applied to answer and resolve many of the open questions and shortcomings of the Big Bang model.

GNM Based Model of the Universe

In Chapter 3, we derived the equation for spontaneous motion of a self-decaying body of mass M_0 at rest (V=0) representing a total relativistic energy, $E_0 = M_0 C^2$. A small portion of the mass spontaneously transforms to energy according to the ESTL as follows:

Transformed energy, $TE = (M_0 - m)C^2$ \hfill (5-10)

This energy is used by the remaining mass m, to propel itself causing a spontaneous motion with a velocity V. The relativistic kinetic energy of the body with the remaining rest mass m and moving at speed V is given by the following equation from ESTL:

Kinetic energy, $KE = m C^2 \left(\dfrac{1}{\sqrt{1-\dfrac{V^2}{C^2}}} - 1 \right)$ \hfill (5-11)

In the absence of any gravitational force or energy, equating this kinetic energy to the energy from mass transformation given by equation (5-10), we obtain the following:

$$(M_0 - m)C^2 = m C^2 \left(\dfrac{1}{\sqrt{1-\dfrac{V^2}{C^2}}} - 1 \right) \quad (5\text{-}12)$$

Simplifying the above provides the following equation, which is same as GNM represented by equation (3-2):

$$m = M_o\sqrt{1-(V/C)^2} \qquad (5\text{-}13)$$

It is to be noted that in the above-described process of mass-energy transformation, the total mass-energy, $E_o = M_o C^2$, is conserved and only the relative mass to energy distribution varies depending upon the velocity ratio V/C or the kinetic energy. This aspect of GNM is different from BBM in that BBM assumes that the total energy remains zero at all times during the evolution of the universe.

As described above, the gravitational effects have been neglected in the above equation, which is a valid assumption when the mass is very small such as for the particles like photons or electrons. However, for the whole universe the total mass M_o is very large and the gravitational effects are dominant especially when the size of the universe is small. In the Big Bang model, equation (5-2), the gravitational potential energy for two masses M and m separated by a distance r is given by:

$$GPE = \frac{GMm}{r} \qquad (5\text{-}14)$$

Applying this equation to a simplified gravitational model of the universe depicted in Figure 5-1, the following integration is obtained for evaluating the gravitational potential energy of the universe:

$$GPE = \int_0^R \frac{Gmm^*}{r}$$

or,
$$GPE = \int_0^R \frac{G\left(\frac{4}{3}\pi r^3 \rho\right)\left(4\pi r^2 \rho dr\right)}{r}$$

or, $$GPE = \frac{3G}{5R}\left(\frac{4}{3}\pi R^3 \rho\right)^2$$

or, $$GPE = \frac{3Gm^2}{5R} \quad (5\text{-}15)$$

Now, from energy balance including the gravitational potential energy (GPE) and the kinetic energy (KE), we get,

$$TE = KE + GPE$$

or, $$(M_0 - m)C^2 = mC^2\left\{\frac{1}{\sqrt{1-\left(\frac{V}{C}\right)^2}} - 1\right\} + \frac{3Gm^2}{5R} \quad (5\text{-}16)$$

Equation (5-16) represents GNM based universe model including the effects of gravity, as opposed to BBM described by equation (5-2). It should be noted that while the total energy E_T in BBM was an unknown and assumed to be zero for a flat universe, the total energy (excluding the mass energy) in GNM based universe model is equal to the transformed mass-energy calculated by equation (5-10).

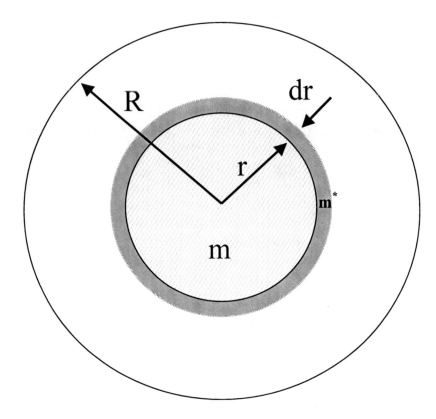

Figure 5.1: Simplified gravity model of the universe.

The Hidden Factor

The above model represents a universe with empty space containing no inherent energy. In BBM, Einstein proposed a 'Cosmological Constant' denoted by Λ, that represents a contribution to the density of the universe from vacuum energy. If Λ is greater than zero, then a spatially flat universe with low matter density can exist due to the contribution from the vacuum energy. An accelerating universe can exist only when Λ is greater than zero. In the case of BBM with a non-zero Λ, equation (5-7), the effect of the Cosmological Constant was explicitly added as a fudge factor to include the contribution of the vacuum energy to the universe expansion. This led to what Einstein described himself as the "Biggest Blunder ...". In GNM universe model, no such extraneous fudge factor exists. It will be shown that GNM model of equation (5-16) closely predicts the behavior of the observed universe without resorting to the phenomena of the incredible inflation caused by extraneous vacuum energy. However, to demonstrate the equivalence and correspondence between the Transformation Energy, TE and the so-called vacuum energy in BBM, we will utilize the following scheme to achieve a closed form solution of GNM-based universe model equation (5-16). Equating the proposed vacuum energy equation by Einstein to the kinetic energy one obtains the following:

$$\frac{1}{6}\Lambda m C^2 R^2 = mC^2 \left\{ \frac{1}{\sqrt{1-\left(\frac{V}{C}\right)^2}} - 1 \right\} \quad (5\text{-}17)$$

or,
$$\Lambda = \frac{6}{R^2} \left\{ \frac{1}{\sqrt{1-\left(\frac{V}{C}\right)^2}} - 1 \right\} \quad (5\text{-}18)$$

Combining equations (5-16) and (5-17) and simplifying results in the following:

$$\Lambda = \frac{6}{R^2}\left\{\left(\frac{M_0}{m} - 1\right) - \frac{3Gm}{5RC^2}\right\} \quad (5\text{-}19)$$

Relativistic Hubble Model

In order to achieve a simplified closed form solution of equations (5-16) through (5-19), we can use the Hubble's Law, V=HR, given by equation (5-1). However, at large values of R, V can exceed the speed of light C and result in singularities in equation (5-16). To avoid this problem, the following alternate equation is proposed to relate the Hubble Velocity V and universe radius R.

$$\frac{H^2 R^2}{2C^2} = \left\{\frac{1}{\sqrt{1-\left(\frac{V}{C}\right)^2}} - 1\right\} \quad (5\text{-}20)$$

This equation can be simplified in following two forms:

$$R = \sqrt{\frac{2C^2}{H^2}\left\{\frac{1}{\sqrt{1-\left(\frac{V}{C}\right)^2}} - 1\right\}} \quad (5\text{-}21)$$

and,

$$\frac{V}{C} = \sqrt{1 - \left\{\frac{1}{1 + \frac{H^2 R^2}{2C^2}}\right\}^2} \qquad (5\text{-}22)$$

Equation (5-22) describes the Relativistic Hubble Model (RHM) as an alternative to the more commonly known Linear Hubble Model $V = HR$. The justification for this relationship is that for the range of observed galactic distances (up to approximately 5 to 9 billion light-years) wherein the linear Hubble's Law is seen to hold, the proposed equation closely matches the predictions of the linear Hubble's Law, as shown in Figure 5-2. The expansion velocity calculated by the Linear Hubble Model exceeds the velocity of light C and hence, violates the theory of relativity for values of R larger than approximately 14 billion light-years. The Relativistic Hubble Model corrects this problem since the predicted V approaches the value of the speed of light as R increases indefinitely. Since the predicted V never exceeds C, the relationship never violates the Einstein's theory of specific relativity and avoids any singularities in the solution to GNM-based universe model given by equation (5-16).

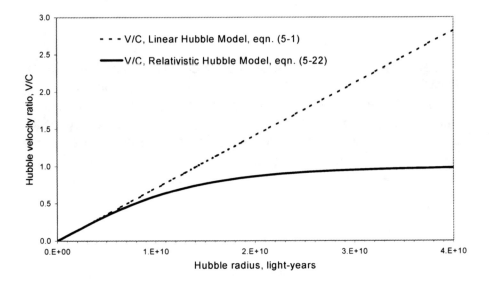

Figure 5-2: Linear versus Relativistic Hubble Model velocity ratio, V/C.

Combining equations (5-18) and (5-20), the following is obtained:

$$\Lambda = \frac{3H^2}{C^2} \qquad (5\text{-}23)$$

It should be noted that for the Relativistic Hubble Model (RHM), the Cosmological Constant Λ given by equations (5-19) and (5-23) is a universal constant depending directly upon the Hubble Constant H and the universal conservation constant C, commonly referred to as the speed of light. However, as will be shown later, for the linear Hubble Model (LHM), V=HR, the Cosmological Constant Λ is not a constant and varies with the size R of the universe.

Critical Mass Density of the Universe

As in the case of BBM, the critical density ratio, ρ_{crit}, equation (5-6), can be obtained equating the kinetic energy from equation (5-17) with the gravitational potential energy, equation (5-15) as follows:

$$\frac{1}{6}\Lambda m C^2 R^2 = \frac{3Gm^2}{5R}$$

and the critical mass density,

$$\rho_{crit} = \frac{m}{\left(\frac{4}{3}\pi R^3\right)} = \frac{\Lambda C^2}{8\pi\left(\frac{3}{5}\right)G} = \frac{5H^2}{8\pi G} \qquad (5\text{-}24)$$

Maximum Mass of the Universe

Now to determine the maximum value of the universe mass m, we can differentiate equation (5-19) with respect to R. This gives:

$$\frac{dm}{dR} = \frac{-\frac{1}{3}\Lambda mC^2 R + \frac{3}{5}\frac{Gm^2}{R^2}}{\left\{C^2\left(1+\frac{\Lambda R^2}{6}\right) + \frac{6}{5}\frac{Gm}{R}\right\}} \qquad (5\text{-}25)$$

Setting this derivative equal to zero provides the relationship between the maximum mass, m_{max}, and the corresponding size of the universe, R_{mmax}, during the expansion of the universe. This gives:

$$-\frac{1}{3}\Lambda m_{max} C^2 R_{m\,max} + \frac{3}{5}\frac{Gm_{max}^2}{R_{m\,max}^2} = 0 \qquad (5\text{-}26)$$

Or,

$$m_{max} = \frac{5}{9}\frac{\Lambda C^2 R^3_{m\,max}}{G} \qquad (5\text{-}27)$$

The mass density of the universe when the maximum mass occurs during the expansion, can now be calculated as follows:

$$\rho_{m\,max} = \frac{m_{max}}{\frac{4}{3}\pi R^3_{m\,max}} = \frac{5}{12}\frac{\Lambda C^2}{\pi G} \qquad (5\text{-}28)$$

Substituting equation (5-23), one can also obtain:

$$\rho_{m\,max} = \frac{5}{4}\frac{H^2}{\pi G} \qquad (5\text{-}29)$$

Comparing equation (5-24) and (5-29),

$$\rho_{m\,max} = \frac{5}{4}\frac{H^2}{\pi G} = 2\rho_{crit} \qquad (5\text{-}30)$$

The size, R_{mmax}, of the universe at which the maximum mass occurs is obtained by substituting m_{max} from equation (5-27) into (5-19). This gives the following equation relating R_{mmax} and M_o :

$$R^5_{m\,max} + \frac{2}{\Lambda} R^3_{m\,max} - \frac{18}{5} \frac{GM_o}{(\Lambda C)^2} = 0 \qquad (5\text{-}31)$$

Equation (5-31) can be solved numerically for a known mass M_o to determine the radius of at which the maximum mass occurs during the expansion of the universe.

Combining equations (5-19) and (5-27), the following is obtained for the maximum mass ratio,

$$\frac{m_{max}}{M_o} = \frac{1}{1 + \left(\dfrac{\Lambda R^2_{m\,max}}{2}\right)} \qquad (5\text{-}32)$$

General Solution of GNM-based Expansion of the Universe

Equation (5-19) represents a quadratic equation that can be solved to obtain mass m of the universe as a function of its size R as follows,

$$m = \frac{5RC^2}{6G}\left[\sqrt{\left(1+\frac{\Lambda R^2}{6}\right)^2 + \frac{12GM_o}{5RC^2}} - \left(1+\frac{\Lambda R^2}{6}\right)\right] \qquad (5\text{-}33)$$

The dark matter mass m_{dm} is generally referred to as the invisible mass in BBM. In GNM, the dark matter mass is defined as the differential mass $(M_o - m)$ that spontaneously decays or converts to energy to compensate for the gravitational energy and kinetic energy of the universe. The dark matter mass is calculated as follows,

$$m_{dm} = M_o - m \qquad (5\text{-}34)$$

Dividing by the total relativistic energy, $E_0 = M_0 C^2$, equation (5-16) can be rewritten in the following non-dimensional form:

$$\frac{mC^2}{E_o} + \frac{mC^2}{E_o}\left\{\frac{1}{\sqrt{1-\left(\frac{V}{C}\right)^2}} - 1\right\} + \frac{3Gm^2}{5RE_0} = 1 \qquad (5\text{-}35)$$

Now, the non-dimensional relativistic mass energy Ω_{ME}, kinetic energy Ω_{KE} and gravitational potential energy Ω_{GPE} are calculated as follows:

$$\Omega_{ME} = \frac{mC^2}{E_0} \qquad (5\text{-}36)$$

$$\Omega_{KE} = \frac{mC^2}{E_o}\left\{\frac{1}{\sqrt{1-\left(\frac{V}{C}\right)^2}} - 1\right\} = \frac{\frac{1}{6}\Lambda mC^2 R^2}{E_0} \qquad (5\text{-}37)$$

$$\Omega_{GPE} = \frac{3Gm^2}{5RE_0} \qquad (5\text{-}38)$$

Substituting into equation (5-35), the following is obtained:

$$\Omega_{ME} + \Omega_{KE} + \Omega_{GPE} = 1$$

The dark matter energy *DME*, is defined in terms of the dark matter mass m_{dm}, as follows:

$$DME = m_{dm}C^2 = (M_0 - m)C^2 \qquad (5\text{-}39)$$

Combining the above with equation (5-16) leads to,

$$DME = mC^2 \left\{ \frac{1}{\sqrt{1-\left(\frac{V}{C}\right)^2}} - 1 \right\} + \frac{3Gm^2}{5R} = KE + GPE \quad (5\text{-}40)$$

Dividing by the total relativistic energy, $E_0 = M_0 C^2$, the non-dimensional dark matter energy is obtained as follows:

$$\Omega_{DME} = \frac{(M_0 - m)C^2}{E_0} = (1 - \Omega_{ME}) = (\Omega_{KE} + \Omega_{GPE}) \quad (5\text{-}41)$$

Results of GNM Based Model of the Universe

In this section, some results of the universe evolution based on GNM described above are presented. Comparison against BBM results and sensitivity results to key parameters and assumptions is also discussed.

The evolution of the universe mass m as a function of its size R is determined using equation (5-33) and input constants measured from recent experiments. Based on recent observational results from two balloon-borne telescopes, Boomerang and MAXIMA [12] the total mass M_o of the universe is estimated to be 100 trillion trillion trillion trillion tonnes or 10^{53} kilograms or 5×10^{22} solar. The recent 2dF Galaxy Redshift Survey [13] designed to measure the redshifts of 250,000 galaxies, recently reported existence of a low-density universe with the Hubble Constant H equal to approximately 70 km sec^{-1} Mpc^{-1} or 2.27×10^{-18} sec^{-1}. Other constants used in calculations are the speed of light, $C = 3 \times 10^8$ m/sec and Gravitational Constant, $G = 6.7 \times 10^{-11}$ m^3/kgm/sec^2. Using the above value of H, the cosmological constant is calculated to be 1.72×10^{-52} m^{-2} from equation (5-23).

Actual and Dark Matter Mass/Energy

Equation (5-33) is solved for the actual mass m of the universe as a function of its size or radius R, as shown in Figures 5-3a,b and c. The radius of the universe is shown on the horizontal axis in units of light-years calculated by dividing the radius R by the speed of light. Hence, the horizontal axis in Figure 5-3a also denotes the time traveled by light from the edge of the universe to the center or in other words the age of the universe in light-years. The actual mass increases with increasing size or age of the universe until a maximum mass is reached at about 9 billion light-years. The mass decreases with size during later years as the universe expands to bigger and bigger sizes. Figure 5-3b and 5-3c depict the same information as Figure 5-3a except that the horizontal axes represent the universe age in seconds and radius in meters respectively rather than light-years. The mass of the universe is less than Planck's mass when the universe age is of the order of 10^{-110} seconds or its radius is of the order of 10^{-100} meters. It is to be noted that at still smaller values of the age or radii, the predicted mass by GNM decreases in direct proportionality with the actual age or radius. GNM thus has no singularity at small values of radius and time of the age of the universe and does not have an exact and absolute time moment of the beginning of the universe similar to BBM. Figure 5.3d shows the ratio of actual to total mass of the universe over a period of one to a trillion light-years.

The mass evolution of the universe is explained from energy considerations as depicted in Figure 5-4, which shows variation of the fractional or non-dimensional mass energy, gravitational potential energy, and kinetic energy described by equations (5-36) through (5-38) above. At smaller age or size, the kinetic energy is small and the universe is dominated by the gravitational potential energy that requires a substantial amount of the total maximum mass M_o to convert to the gravitational energy leading to a decreased mass of the universe. As the size of the universe increases, the gravitational energy decreases and the kinetic energy increases. At R greater than 100 billion light-years, most of the energy of the universe is in the form of kinetic energy with both the mass energy and gravitational potential energy diminishing to small values. The total fractional gravitational and kinetic energy, also defined as the fractional dark matter energy, equation (5-41), is also shown in Figure 5-4. The dark matter energy dominates at small sizes. As the size R increases, it first decreases and then increases with a minimum occurring at approximately 9 billion years which coincides with the time when the maximum universe mass occurs.

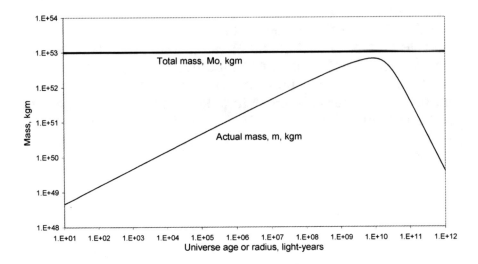

Figure 5-3a: Universe mass versus age or size predicted by GNM.

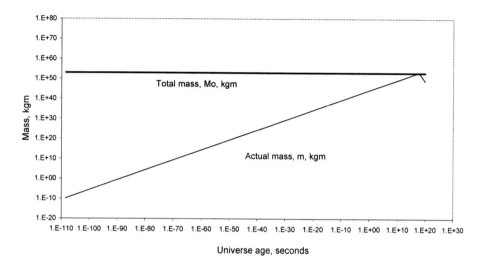

Figure 5-3b: Universe mass versus age in seconds predicted by GNM.

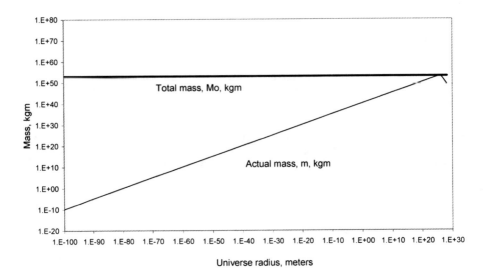

Figure 5-3c: Universe mass versus size in meters predicted by GNM.

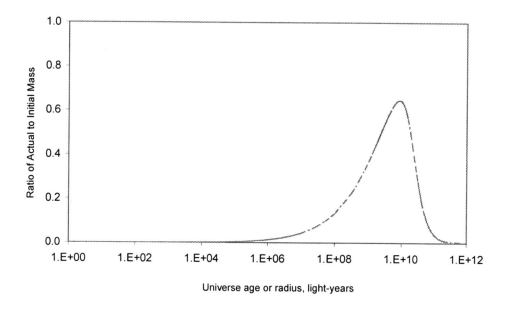

Figure 5-3d: Universe mass ratio versus age or size predicted by GNM.

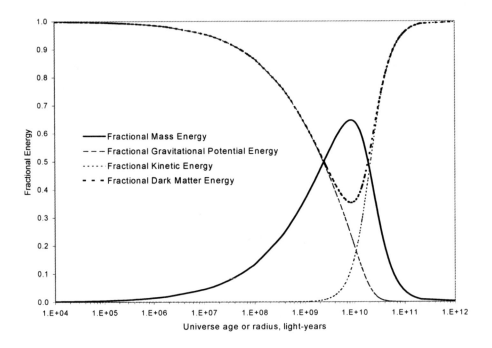

Figure 5-4: Fractional Mass Energy, Gravitational Energy and Kinetic Energy versus universe age or size predicted by GNM.

The Hidden Factor

Figure 5-5 shows the ratio of the actual mass density to the mass density corresponding to the maximum mass, ρ_{mmax}, of the universe. It should be noticed that this density ratio is very large when the universe is small in size and decreases as the universe grows. The mass of the universe increases so long as the actual mass density is greater than ρ_{mmax} until the universe is approximately 9 billion years old. At later times, the actual mass density continues to decrease to values smaller than ρ_{mmax} leading to an accelerated expansion and decrease in mass as it grows larger and larger. Figure 5-6 shows the mass and density evolution of the universe on a smaller size and time scale. It should be noted that in spite of a large decrease in mass at the very small universe size below even Planck's scale (10^{-35} meters), the density remains very large due to the excessively small volume. Also shown in Figure 5-6 is the ratio of dark matter mass to actual mass. As the universe size increases, this ratio decreases from a very large value ($>10^{30}$) to the order of one at the age of 9 billion years. As opposed to BBM, which requires the mass density to remain equal to the critical mass density throughout the evolution of the universe, GNM predicted mass density decreases with increasing size or age of the universe.

Equation (5-31) is solved numerically for a known initial mass M_0 to determine the radius corresponding to the occurrence of the maximum mass during the expansion of the universe. Equation (5-32) is used to calculate the corresponding maximum mass to the total mass ratio. The predicted results are shown in Figure 5-7. As the mass M_0 increases, the gravitational effects become more dominant and the universe age or size at which the maximum mass occurs increases. For mass M_0 less than approximately 10^{20} solar, the maximum mass ratio is always equal to one i.e. the maximum mass of the universe is equal to the total mass. This is due to the fact that for smaller total mass, the dark matter energy or the sum of gravitational and kinetic energies becomes significantly (several orders of magnitude) less than the total mass energy $E_0 = M_0 c^2$ at some intermediate sizes of the universe. The universe age or radius at which it occurs is approximately 4 to 5 billion light-years. For mass M_o greater than approximately 10^{22} solar, the maximum mass ratio of the universe is less than one and occurs at an age or size larger than approximately 5 billion years. This is shown in Figure 5-8 wherein the actual to initial mass ratio and the maximum mass ratio is shown as a function of the universe age or size for three different total mass values ranging from 10^{18} solar to 5×10^{22} solar. For mass values of 10^{18} solar and 10^{21} solar the maximum mass ratios are 1 and 0.95 respectively. For M_o equal to 5×10^{22} solar, the maximum mass ratio is approximately 0.7.

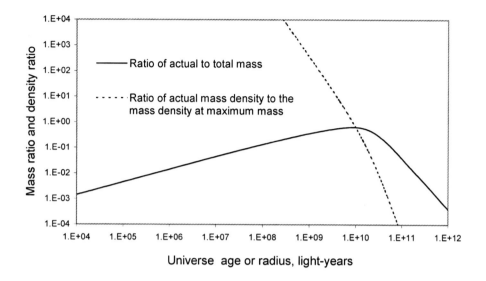

Figure 5-5: Ratio of actual mass density to the mass density ρ_{mmax} and mass fraction as function of the universe age or radius.

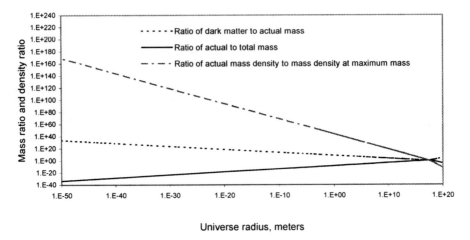

Figure 5-6: Ratio of actual mass density to mass density ρ_{mmax} and mass fraction versus universe size at small scale.

The Hidden Factor

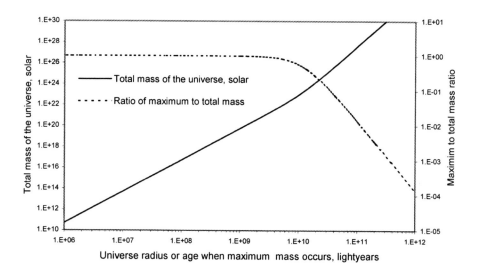

Figure 5-7: Ratio of maximum to total mass and total mass versus universe age or radius at which the maximum mass occurs.

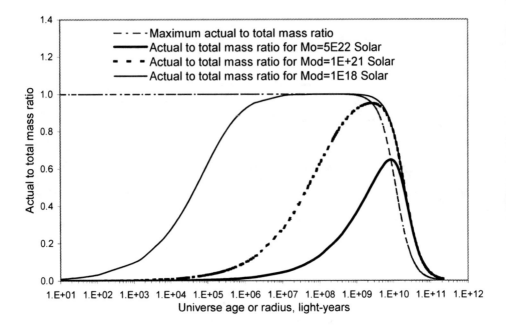

Figure 5-8: Ratio of maximum to total mass as a function of total mass versus universe age or size.

The Hidden Factor

Linear versus Relativistic Hubble Model

Most widely used theories of the Big Bang model (BBM), employ the Linear Hubble Model (LHM), *V=HR*, to describe the observed universe. As discussed earlier in this chapter, the expansion velocity calculated by the Linear Hubble Model based on the currently observed values of the Hubble Constant *H*, exceeds the velocity of light at a radius or age of the universe that exceeds about 14 billion years. No visible objects such as stars, galaxies or supernovas have been observed at such great distances. The reason given for this is that the light from such objects can not reach us because they are beyond the visible horizon. Another postulate is that such a distance denotes the early phase of the birth of the universe when no matter or objects were yet formed. GNM employs the Relativistic Hubble Model (RHM) given by equation (5-22) to eliminate shortcomings of the LHM at large radii while maintaining the observed linear relationship of the LHM at radii smaller than a few billion light-years. The impact of the Linear versus Relativistic Hubble Model on the evolution of mass, dark matter and expansion of the universe using GNM is discussed below.

Figure 5-9 shows GNM predicted velocity ratio, V/C, and mass ratio of actual to total mass for the Linear versus Relativistic Hubble models. For the LHM, as the velocity increases, the mass first increases and reaches a maximum when the universe is about 9 billion light-years. Beyond this point, the mass deceases and finally becomes zero, as V equals the speed of light C at the age of about 14 billion light-years. This is consistent with BBM supported notion that the birth of the universe occurred at about 14 billion years ago. The predictions of the RHM coincide with the LHM up to about 5 billion light-years. However, the RHM predicts a continuous evolution with a continuously increasing velocity and decreasing but a significant amount of mass existing beyond 14 billion light-years. The prediction of the RHM thus negates existence of an absolute time of the beginning of the universe or a limited visible horizon. RHM does not support a universe with a finite age involving an absolute time of the beginning (the Big Bang) or ending (the Big Crunch). Figure 5-10 shows the dark matter to total mass ratio and the ratio of dark matter mass to actual (visible) mass for the two models. Again, for LHM, the dark matter mass begins to increase much more rapidly than the RHM for radii greater than 5 billion light-years, and all energy exists as dark matter at R greater than 14 billion light-years. Predictions of both LHM and RHM match closely for universe radii or age less than approximately 5 billion light-years.

The impact of LHM versus RHM on the Cosmological Constant Λ is shown in Figure 5-11. It should be noted that the Cosmological Constant for the RHM is given by equation (5-23) and remains invariable for all sizes or ages of the universe. However, the Cosmological Constant for the LHM given by equation (5-18) remains constant only for universe age or size up to about 2 billion years as shown in Figure 5-11 and increases exponentially to very large values as the universe increases in size or age. Hence, the assumption of a non-varying universal Cosmological Constant is not consistent with the LHM.

Impact of the Hubble Constant

The impact of the Hubble Constant H on the Hubble velocity is shown in Figure 5-12 wherein the Hubble Constant is varied in the range from 50 km sec^{-1} Mpc^{-1} (1.64×10^{-18} sec^{-1}) to 80 km sec^{-1} Mpc^{-1} (2.6×10^{-18} sec^{-1}). As expected, smaller the Hubble Constant, longer it takes for the velocity to attain a value equal to C for both the LHM and RHM. The effect of varying H on the predicted mass ratio for the RHM is shown in Figure 5-13. For larger H, the maximum mass occurs at approximately 8 billion light-years versus 12 billion light-years for the smaller H. In other words, if the current time corresponds to the maximum mass, the age of the universe would be approximately 4 billion years older for the smaller H as compared to the larger H.

Evolution of Large Scale Structures

GNM calculates an overall mass, shown in Figure 5-3, of the universe as a function of its age or size. The actual universe contains a population of stars, galaxies, clusters and super-clusters. A first order and very preliminary distribution of these during the evolution of the universe can be obtained as follows. Assuming that each star weighs 2×10^{30} kilograms (the mass of our sun), a galaxy consists of a minimum of a thousand billion (1×10^{12}) stars, a cluster consists of a minimum of a million (1×10^{6}) galaxies and a super-cluster contains a minimum of a thousand (1×10^{3}) clusters, Figure 5-14 shows the corresponding first order approximation of the population of stars and structures as a function of the universe age or size.

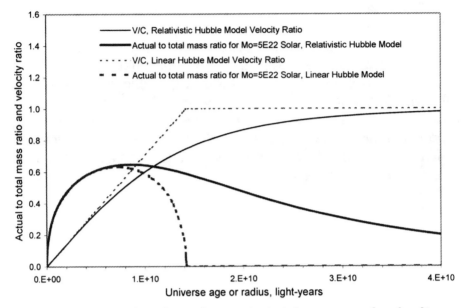

Figure 5-9: Ratio of maximum to total mass and velocity ratio for Linear Hubble Model versus Relativistic Hubble Model.

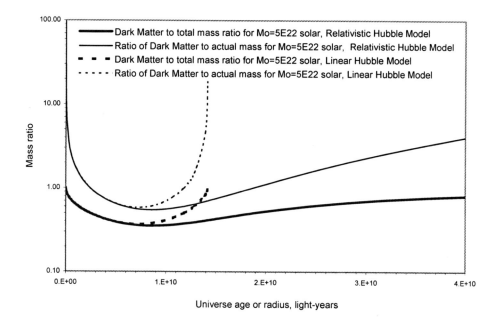

Figure 5-10: Ratios of dark matter to total and actual mass for Linear versus Relativistic Hubble Model.

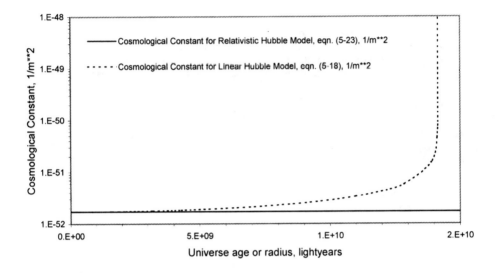

Figure 5-11: Cosmological Constant for Linear and Relativistic Hubble Models.

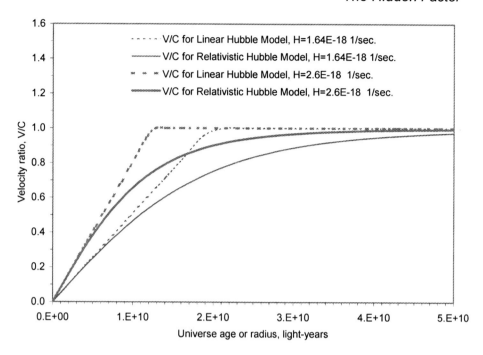

Figure 5-12: Sensitivity of Linear and Relativistic Hubble Models to the Hubble Constant.

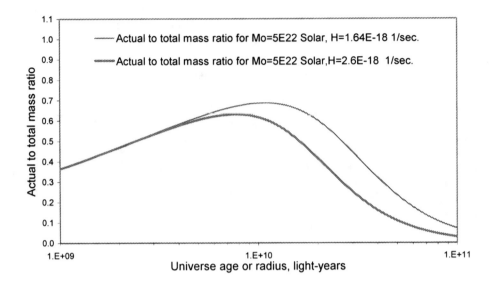

Figure 5.13: Impact of Hubble Constant on actual to total mass ratio for the Relativistic Hubble Model.

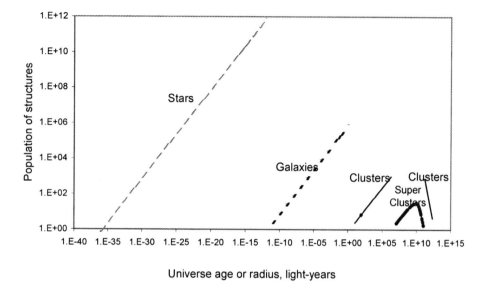

Figure 5-14: First order approximation of the population of stars, galaxies and structures versus universe age or size.

Avtar Singh

GNM resolves the Shortfalls of the Big Bang Model (BBM)

Earlier in this chapter, we described the strengths and weaknesses of the Big Bang Model. We will now discuss how most of the weaknesses are eliminated by GNM while matching closely the predictions of the currently observed universe. Since many of the weaknesses of the Big Bang Model (BBM) relate to the formulation and treatment of the physical phenomena related to mass-energy-space-time, we will first look into the basic models and assumptions in GNM and BBM and their impact on the universe evolution.

Conservation of Mass, Energy and Momentum

In BBM, all the matter and energy in the universe is assumed to emerge from almost nothing. The initial conditions at the beginning of the universe cannot be defined due a singularity in the equations. This physical and mathematical difficulty is swept under the rug of the incredible scenario of inflation that seems to occur at Planck's time of 10^{-35} seconds, which represents the minimum measurable time scale per quantum rules. In the inflation era, the universe is assumed to be filled with a so-called false vacuum that has a large negative pressure. Essentially all the energy of the universe is created in the form of a non-gravitational energy during an exponential expansion of this false vacuum that occurs at Planck's time scale. As the universe expands at an accelerated pace this energy decreases drastically due to the deceasing pressure of the hot gas. The universe also cools down as it expands and eventually this energy transforms during the phase transition to become the visible matter in the universe in the form of planets, stars and galaxies including all life in it. Inflationary scenario thus provides an incredible history of creation from almost nothing at the so-called beginning of time. As the matter formed into small and large scale structures of the universe, some of the early non-gravitational energy transformed into the gravitational energy due the attractive forces among the

The Hidden Factor

material structures in the universe. This gravitational energy is considered to be negative and equal to the remaining positive non-gravitational energy, the two canceling each other resulting in a net zero energy in the universe. All this incredible evolution of the universe from nothing is often characterized by some as the ultimate free lunch. In summary, there is no net energy in BBM universe to conserve. As far as conservation of momentum is concerned, there is no consideration given.

The GNM, on the other hand, is strictly based on the conservation of mass-energy and momentum. The universe is assumed to consist of a fixed total amount of mass-energy given by $E_o = M_o\ C^2$. This total energy is distributed in three components- mass energy ($m\ C^2$), kinetic energy given by equation (5-11) and gravitational potential energy given by equation (5-15). As the universe evolves, the net value of the total energy remains constant while the individual values of the three distributed energies change with the size of the universe as shown earlier in Figure 5-4. The sum of the gravitational and kinetic energies, which are invisible, is referred to as the dark matter energy which is also equal to the total energy minus the mass (visible matter) energy. Thus the state of the universe is fully defined by its total energy and size, with no other initial or boundary conditions required. The momentum conservation in GNM was described in Chapter 3 (Figures 3-2 and 3-3), which is consistent with the observed motion of light and Hubble expansion of the universe.

Now we will look into various specific shortcomings of BBM and how GNM addresses them.

Singularity in BBM initial conditions

The BBM describes a dynamic evolution of the universe as an explicit function of time and physical parameters. The numerical values of the Big Bang model parameters at the beginning of time (t=0) can not be specified because the physical laws considered in GNM break down at the origin of time. The initial conditions must instead be inferred from extremely large extrapolations of the currently observed behavior of the universe back in time. The extrapolation itself suffers from the inherent weaknesses of the model. The history and evolution of the universe up to modern times strongly depends upon the initial conditions, and since they are not known the whole evolution of the universe essentially becomes hypothetical without any help from BBM.

GNM has no need for specified initial conditions since the solution of GNM is strictly based on energy considerations and not time dependent parameters. Time is only an implicit and not an explicit parameter in GNM solution and is inferred by dividing the size or radius of the universe by the speed of light.

Another key question that remains unanswered in BBM is what set the universe in motion? The cosmological expansion is the generic feature of the Einstein's Theory of general relativity and the Newtonian approximations depicted by the equations (5-1) thru (5-9) above. The motion is thus enforced by prescription of governing equations without a solid explanation of the mechanistic physics behind it. The universe is set in motion by the incredible and hypothetical inflation scenario described earlier in this chapter.

In GNM, the expansion energy comes from the spontaneous and relativistic conversion of the existing total energy $E_0 = M_0 C^2$. As discussed earlier in chapter 2,3 and 4, such spontaneity is inherent in nature as observed in the spontaneous decay of small unstable particles as well as in the well-established wave-particle nature of quantum particles. There is no trigger or beginning of time (t=0) condition required for initiation of expansion as it is built in the Cosmological Constant and the Hubble Expansion velocity model, which is entirely based upon the observed data. Hence, GNM eliminates the need for an incredible inflation model as a savior from the critically needed but un-specifiable initial conditions in BBM.

Uncertainty in the historical and current values of parameters

The Big Bang model relies on the values of three parameters (Ω_k, Ω_m and Ω_Λ) described in equation (5-9), to calculate evolution of the universe. The uncertainty induced by the scatter and varied interpretation of the available data to quantify these parameters of the model remains very large reducing the confidence in the predictions of the model. In the current solutions of the Big Bang model, for reasons of simplicity and non-availability of the historical data on the early universe expansion, these parameters are assumed to remain constant during the evolution of the universe from the occurrence of the Big Bang to the current time period. There is no solid evidence available from the observations for supporting or refuting this leap of faith. Even if these parameters were known to vary with time over the evolution of the universe, it would be impossible to define or quantify that variation with the available observations today.

The GNM does not involve physical parameters that can not be directly measured from observations or analytically derived. The two parameters required by GNM, E_o (or M_o) and H have been measured via direct observations without any need for speculation or theoretical derivation.

Unavoidable but Incredible Inflation Scenario

As discussed earlier, even though the Big Bang model explained several aspects of the observed universe, a large coincidence and fine-tuning via an incredible inflation scenario is still required to explain the following three major problems.

The Horizon Problem

Because of the limitations of the speed of light, there would be regions in space, which will remain invisible to an observer because the light from the origin has not arrived as yet in those regions. This creates the horizon problem since such a restriction does not seem to apply to the microwave background radiation, which is observed to be incredibly homogeneous and uniform in temperatures and density without any indication of inaccessible regions in the universe. The incredible homogeneity observed in the background radiation to an accuracy of 3 parts in 100,000 creates an astronomical mystery that fails an appealing explanation. BBM needs to be supplemented by another physical model that would allow enhanced communication (at speeds which are several times faster than the speed of light) among various early and later regions of the universe. The incredible inflation model filled in such a need. In the absence of any other model the scientific community thus far had no other choice.

Other than the communication at almost infinite speed, assumed in the inflation model, the horizon effect prevents any other mechanistic and physical explanation of such a remarkable homogeneity in the properties of the universe with regard to the distribution of galaxies and the cosmic background radiation. GNM explains this observed homogeneity and resolves the horizon problem without the violation of the speed limit set by C as discussed below.

We derived GNM equation (5-33) for predicting the universe mass m as a function of its size using mass-energy balance. For sake of simplicity and to achieve an overall energy balance, the mass of the universe was lumped into one chunk as m, which was assumed to move at an overall lumped velocity V. As discussed in Chapter 3, in addition to satisfying the law of conservation of mass-energy, the moving mass m must also satisfy the law of conservation of momentum. This can only be achieved if, during its decay, the body of mass m breaks up into at least two or more (even number) of pieces that move in opposite directions such that the sum of each individual mass-energy equals the total mass-energy, $E_o=M_oC^2$ and the net sum of momentum (mV) of each individual mass equals zero. Figure 3-1 shows a typical simple schematic of such a process, wherein a body of rest mass M_o decays into two smaller masses m_1 and m_2 moving at velocities V_1 and V_2 respectively.

Another example of such motion is shown in Figure 3-2 in which the original mass breaks into four equal masses that move in symmetrically opposite directions to satisfy the momentum conservation law. Note that it is not necessary that the broken masses are all equal to each other. So long as there are at least two masses that move opposite to each other with equal and opposite momentum (mV), there could be different amounts of paired masses that can conserve the momentum. This can be represented mathematically, as in equation (3-6a):

$$mV = m_1V_1 + m_2V_2 + m_3V_3 + \ldots + m_nV_n = \sum_1^n m_nV_n \quad (5\text{-}42a)$$

Also note that the overall mass-energy ($E_o = M_o C^2$) has to be conserved during this decay process such that the following is satisfied:

$$M_o C^2 = \sum_1^n \frac{m_n C^2}{\sqrt{1-(V_n/C)^2}} \quad (5\text{-}42b)$$

In the above equations each term $m_n V_n$ represents a pair of masses moving with equal and opposite momentum relative to the fixed frame of reference and wherein m_n can have values between zero and m and V_n can vary between zero and C. Similarly, the overall lumped kinetic energy term in equation (5-11) can be written in terms of the distributed masses as follows:

$$\text{Kinetic energy, } KE = \sum_1^n m_n C^2 \left(\frac{1}{\sqrt{1-\frac{V_n^2}{C^2}}} - 1 \right) \quad (5\text{-}43)$$

A large number of mass-velocity configurations are possible that can satisfy equation (5-42). One set of such two dimensional configurations is shown in Figures 3-2. When V equals C, the mass m_n disintegrates or converts completely into energy as depicted in the bottom of Figure 3-2. The scenario can easily be extended to three dimensions, wherein the mass m gets de-fragmented and dissolved into paired fragments moving radially outwards into space at speed V, which could be as small as zero or approach C for some mass fragments.

Now we are ready to discuss how the above detailed description of the universe mass and velocity distribution helps explain the Horizon Problem. As described in Figure 2-4 in Chapter 2, for the decaying mass that moves at velocities close to C, the space/time dilation leads to an effective speed that could be much greater than the speed of light. As postulated during BBM evolution of the universe, that the mass-energy that separated to form the microwave background radiation would essentially behave as if the effective speed attained may be several orders of magnitude larger than C. Due to the space dilation the discontinuities are greatly smeared in space leading to the incredible homogeneity observed in the background radiation to an accuracy of 3 parts in 100,000. This effect is shown in Figure 5-15 wherein the space dilation of this order is predicted by the Einstein's specific theory of relativity given by equations (2-1) and (2-2), when V approaches C at very large distances corresponding to the microwave background radiation. Other physical explanation for the space-time dilation and resulting non-locality were described in chapter 3 and 4 based on the infinite wavelength achieved by quantum particles moving at velocities close to C. It should be mentioned here that while GNM predicts the microwave background radiation homogeneity by space/time dilation at large distances, the possibilities of large masses as stars, galaxies etc. represented by very small V at smaller distances is also predicted by GNM.

In summary, GNM explains the absence of the so-called Horizon Problem and supports the observed homogeneity in the universe naturally without resorting to the incredible phenomenon of inflation.

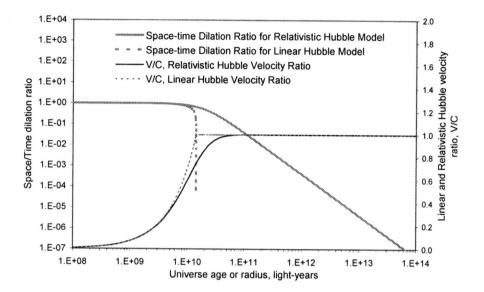

Figure 5-15: Space/time dilation and velocity ratio versus universe age or radius for Linear and Relativistic Hubble Models.

The Flatness Problem

Observations of the current universe show that it is flat without any significant positive or negative curvature. Since both the mass as well as the vacuum energy of the Cosmological Constant contribute to the curvature of the universe, a flat universe implies that the sum of Ω_m and Ω_Λ is close to 1. The flatness problem arises from the incredible fact that this sum is close to 1 when it could theoretically be any number different from 1. The solution of the Big Bang model equations would predict a stable flat universe only if it was flat initially at t=0. This requires the initial condition of flatness needs to be imposed on BBM to match the observed behavior of the universe. In addition, BBM requires the universe to be flat throughout its entire life.

GNM does not require an imposition of flatness or any other restrictions other than the laws of conservation during the evolution of the universe. The flatness problem is specific to BBM formulation and not a physical problem in the real sense. Flatness assumption allows an easy escape from the irresolvable problem of quantifying the total energy E_T, which is assumed to be zero in BBM equation (5-2).

In summary, GNM does not involve the Flatness Problem and does not have to resort to the incredible phenomenon of inflation and an artificial enforcement of the assumption of the critical density at the beginning and throughout the evolution of the universe.

The Cosmological Constant Problem

The GNM provides a physical model for the Cosmological Constant which replaces the current understanding of the Cosmological Constant as the existence of a negative pressure caused by the energy of the vacuum whose sole function is to push the space radially outwards causing it to expand.

The observations of the universe today strongly support a non-zero Cosmological Constant along with a flat universe. As the universe expands, the gravity force diminishes due to larger distances and the Cosmological Constant dominates the accelerating expansion. Before GNM there was no credible physical theory that can predict even the right order of magnitude of the vacuum energy. Particle-physics theories have been put forward that provide only partial explanation for existence of vacuum energy and a non-zero Cosmological Constant, but the values predicted by these are about 120 orders of magnitude greater than the recent supernova observations.

The GNM also answers other unresolved questions regarding the Cosmological Constant. One question is whether the Cosmological Constant is really a universal constant that does not change with the time evolution of the universe. Recently, some scientists have proposed a variable (Quintessence [26]) Cosmological Constant to account for the behavior of the observed data. GNM predicts a universal value for the Cosmological Constant given by equation (5-23).

Another puzzle solved by GNM and that has been haunting scientists is why it comes to dominate at about the epoch of galaxy and star formation and not earlier. This is counter-intuitive to the notion that prime mover for galaxy and star formation is the pull of gravity rather than the expansion induced by anti-gravity signified by the Cosmological Constant. Recently [15], the Anthropic Principle, which has a low approval rating among physicists, has been shown to be the only approach that can naturally resolve the problems related to the Cosmological Constant. GNM predicts the epoch, lasting up to 10 billion light-years as shown in Figure 5-14, of stars and galaxy formation coincident with an equal dominance of the kinetic energy fueled by the Cosmological Constant and the gravitational potential energy. GNM predictions show that the maximum mass in the universe occurs when the gravitational energy is in balance with the kinetic energy.

In summary, GNM provides a credible physical model that answers these serious outstanding puzzles related to BBM without the need of resorting to the incredible and unverifiable inflation scenario. At the same time GNM eliminates the singularities and irresolvable problems of specifying initial conditions that inflict BBM at early times of the evolution. GNM also restores the long-standing confidence in the laws of conservation of mass-energy and momentum, without the need for creating a universe out of nothing, and buying into an incredible assumption that total energy of the universe is zero. GNM provides a model that does not need a savior in inflation. GNM also diminishes the need for the Anthropic Principle, which is needed and hence, invented as a last resort to resolve in a half-haphazard and far-fetched way, the problems of the Cosmological Constant.

Comparison of GNM against Recent Astronomical Observations

Puzzle of Dark Matter or Dark Energy

Based on recent observations [5] on the intensity of light from the most distant exploding star called type 1A supernova, astronomers have confirmed that the expansion of the universe is accelerating at a faster and faster rate. This is believed to be caused by an anti-gravity force or the so-called dark matter energy pushing galaxies apart. These results challenge BBM, which predicts that the pull of gravity would slow down the expansion rate of the universe as it evolves from its beginning. The new observations suggest that such a scenario could have happened only during the first few billion years of the age of the universe. The later expansion was dominated by the anti-gravity energy causing the observed run-away expansion.

The type 1A supernovas are believed to possess the same intrinsic brightness and hence, are used as standard candles. Astronomers observe the intensity and distance of these exploding stars, as they appeared several billion years ago. If the universe expansion was decelerating due to gravity, the distance between the earth and these stars would be smaller and their intensity would be brighter than if the universe expansion was accelerating. The recent observations by the Hubble Space Telescope of these distant stars located as far as 10 billion years from us have confirmed that their intensity is about 20% to 100% dimmer than expected for a gravity-dominated universe predicted by the classic BBM. These observations consistently confirm the ever-increasing rate of universe expansion during the later history of the universe. Scientists have ruled out other possible causes for such lower dimness, such as dust and varying intrinsic brightness.

The BBM lacks appropriate physical models to account for the accelerated expansion of the universe. A non-zero Cosmological Constant is included as an empirical fudge factor to correct this problem and to match the observed Hubble expansion. Predictions of GNM shown in Figure 5-4 provides a physical explanation of the dark matter energy that causes the observed run-away expansion of the universe. The dark matter energy during the early age, up to about 2 billion years, of the universe consists primarily of the gravitational potential energy. At about 9 billion years, the gravitational and the anti-gravitational or kinetic energy even out. Following this period, the anti-gravitational energy in the form of the increasing kinetic energy dominates the universe expansion. Figure 5-2 shows the accelerating universe expansion predicted by GNM Relativistic Hubble Model (RHM) wherein the radius of the universe increases at an exponentially faster rate for a given velocity ratio as compared to the LHM used in the standard BBM.

Observation of High Redshift Objects with Lower Redshift Galaxies

Since the 1980's, the high-energy radiation X-ray, gamma ray and ultra-high cosmic ray data has been collected from objects in the local super cluster [16]. Based on this data, the observers conclude that a large and old galaxy ejects new material, which may form younger and smaller companion galaxies around it. The younger galaxies may further act as a source of ejected material giving rise to the formation of still younger quasars and objects. The age hierarchy is observed to exist from the characteristics of the paired groups of galaxies across and their luminous connections to the active nuclei and from the emitted radio and higher energy signals. A strong correlation between the age and redshift of the ejected objects is seen; younger the material the higher is the observed redshift. The overall conclusions of these observations described in ref. [16] are as follows:

1. A large number of the observed quasars associated with much lower redshift galaxies provide evidence that quasars are energized condensations of matter, which have been ejected from active galaxy nuclei.
2. Objects, which appear young, are aligned on either side of eruptive objects. This implies ejection of protogalaxies.
3. The youngest objects appear to have the highest redshifts. This implies that intrinsic redshift decreases as the object ages.
4. As the distance from the ejecting central object increases, the quasars increase in brightness and decrease in redshift. This implies that the ejected objects evolve as they travel outward.
5. At a redshift of about 0.3 and a distance of about 400kpc from the parent galaxy the quasars appear to become very bright in optical and X-ray luminosity. Clusters of galaxies, many of which are strong X-ray sources, tend to appear at these distances.
6. Clusters of galaxies in the range of redshift from 0.2 to 0.4 contain blue, active galaxies. It implies that the galaxies continue to evolve to higher luminosity and lower redshift in the range of 0.01 to 0.2, such as Abel clusters that lie along ejection lines from galaxies like CenA.
7. The strings of galaxies, which are aligned through the brightest nearby spirals, have redshifts of 0.01 to 0.02. Presumably they are the last evolutionary stage of the

ejected protogalaxies before they become slightly higher redshift companions of the original ejecting galaxies.

In summary, newly created, high redshift material is ejected in opposite directions from active galaxies. The material evolves into high redshift quasars and then into progressively lower redshift objects and finally into normal galaxies. As these newer galaxies age, and grow in size and mass, they in turn eject newer generations evolving into clusters of large number of galaxies in a cascading process. These observations are counter to the conventional view of BBM that the galaxy formation occurs via condensation out of some tenuous, homogeneously pervading hot gas.

We will now evaluate these data using GNM and discuss how some of the not so conventional phenomena observed in these data can be explained and predicted by GNM. These phenomena include creation of matter, paired and opposite ejections from the parent source and impact on the observed redshifts as well as the Hubble Constant.

Creation or Formation of Matter

The standard BMM assumes that all the matter and energy in the universe emerges from almost nothing. Essentially all the energy of the universe is created in the form of non-gravitational energy during an exponential expansion of the false vacuum that occurs at Planck's time. As the universe expands at an accelerated pace this energy decreases drastically due to the deceasing pressure of the hot gas. The universe also cools down as it expands and eventually this energy transforms during the phase transition to become the visible matter in the universe in the form of galaxies, stars and planets including all life in it. Inflationary scenario thus provides an incredible history of creation from almost nothing at the so-called beginning of time. As the matter formed into small and large scale structures of the universe, some of the early non-gravitational energy transformed into the gravitational energy due the attractive forces among the material structures in the universe.

The GNM, on the other hand, predicts formation and distribution of matter in the universe strictly based on the conservation of mass-energy and momentum. The universe is assumed to consist of a fixed total amount of mass-energy given by $E_o = M_o C^2$. This total energy is distributed in three components- mass energy ($m C^2$), kinetic energy given by equation (5-11) and gravitational potential energy given by equation (5-15). As the universe evolves, the net value of the total energy remains constant while the individual values of the three distributed energies change with the size of the universe as shown in Figure 5-4. The sum of the gravitational and kinetic energies, which are invisible, is referred to as the dark matter energy which is also equal to the total energy minus the mass (visible matter) energy. Thus the creation of matter is via transformation of the dark matter energy into the visible or actual mass energy. This is shown in Figure 5-16, wherein the fractional dark matter energy and actual mass energy are calculated using GNM for a galaxy with an initial mass M_0 of 1×10^{12} solar. When the galaxy is less than 10^{-8} kiloparsec in size, the formation of matter occurs when the dark matter energy consisting mainly of the gravitational potential energy dominates its evolution. As the size of the galaxy increases, the dark matter energy entirely converts to the mass energy at about 0.001 kiloparsec in size. This creation of matter predicted by GNM is consistent with the observations [16]. Beyond about 1 million kiloparsec of age, the kinetic energy generated via conversion of the mass energy dominates the expansion of the galaxy.

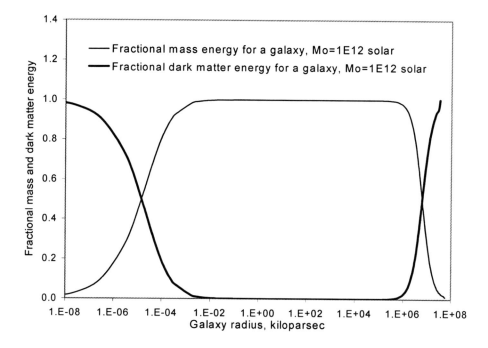

Figure 5-16: Fractional mass energy and dark matter energy of a galaxy as a function of its radius or age.

Paired and Opposite Ejection of Created Matter

We derived GNM equation (5-33) for predicting the mass m as a function of the galaxy or universe size using mass-energy balance. For sake of simplicity and to achieve an overall energy balance, the mass of the universe was lumped into one chunk as m, which was assumed to move at an overall lumped velocity V. As discussed in Chapter 3, in addition to satisfying the law of conservation of mass-energy, the moving mass m must also satisfy the law of conservation of momentum. For a body at rest (V=0) initially, the net momentum is zero. This momentum can be maintained at zero if the body of mass m breaks up into at least two or more (even number) of pieces that move at equal speed V in opposite directions. An example of such motion is shown in Figure 3-2 in which the original mass breaks into four equal masses that move in symmetrically opposite directions to satisfy the momentum conservation law. Please note that it is not necessary that the broken masses are all equal to each other. So long as there are at least two equal masses that move opposite to each other with equal and opposite magnitudes of momentum (mV), there could be different amounts of paired masses that can conserve the momentum of the overall system or the universe. This is represented mathematically as described in equations (5-42) and (5-43), and related discussion above.

A large number of mass-velocity configurations are possible that can satisfy equation (5-42). One set of such two dimensional configurations is shown in Figures 3-2. As the galaxy expands, its mass density decreases because of the increasing volume. When V equals C, the mass disintegrates or converts completely into energy as depicted in the bottom of Figure 3-2. The scenario can easily be extended to three dimensions, wherein the mass m gets de-fragmented and dissolved into paired fragments moving radially outwards into space at speed V, which could be as small as zero or approach C for some mass fragments.

The Hidden Factor

Again, GNM provides a physical basis for the observed paired and opposite ejections of the created matter in several galaxies in reference [16]. Before we analyze the impact on redshifts of the ejecting matter as predicted by GNM, we will discuss the observed and predicted rotational velocities of stars within galaxies. Once we understand the distribution of the rotational as well as radial velocities within and around a galaxy, it will be easier to explain the observed redshifts behavior as described above.

Rotational and Radial Velocities in Galaxies

Stars in the spiral galaxies are observed to rotate with a finite tangential velocity around the center of the galaxy due to the attractive pull of the gravity of the matter. The observed tangential velocities of the stars are so fast that the centrifugal force of rotation ought to make them fly off into the intergalactic space. The astronomers have, until now, explained such large rotation velocities by claiming existence of large amounts of invisible dark matter. The existing dark matter theories suggest that the dark matter may exist in the form of black holes, unknown subatomic particles, dead or dwarf stars or planets in a massive "halo" around the center of the galaxy and extending out radially outward to the outer boundaries of the galaxy. By equating the net centrifugal force of rotation and the attractive force of gravity, the tangential velocity of rotation can be calculated as follows:

$$V_T = \sqrt{\frac{Gm}{R}} \qquad (5\text{-}44)$$

The rotational velocity calculated from the above formula depends upon how the mass m varies with the radius or size of the galaxy. Among many possibilities, three specific cases are considered below:

1. Concentrated mass at the center (Black Hole):

In this case it is assumed that all the mass of the galaxy is concentrated at the center of the galaxy in a spheroid, known as black hole, just as the mass in a solar system is concentrated in the Sun. Hence, m would remain constant and equal to the total mass of the galaxy for all radii greater than the outer radius of the black hole. The rotational velocity would then vary as the inverse of the square root of the radial distance from the center of the galaxy, which is also known as the Keplerian decrease.

2. Constant density distribution:
In this case the mass increases linearly with the volume or as the cube of the radial distance from the center of the galaxy resulting in a linear increase in the rotational velocity similar to a solid body rotation.

3. Mass m increases linearly with R
If m increases linearly as the distance form the center of the galaxy, equation (5-44) results in a constant rotational velocity irrespective of the distance from the center.

Observed rotation curves of spiral galaxies and implications of dark matter

The measured or observed rotation curves of spiral galaxies provide a tool for evaluating the mass distribution of stars in galaxies, which in turn can be used to derive understanding and details of galaxy evolution. Numerous investigations [17 through 25] have been performed to measure the rotation velocities of galaxies of different sizes and shapes using several different measurement techniques. Rotational velocity data from numerous observed galaxies are available on the World Wide Web. The measurement techniques have evolved over the past decades with improving accuracy and resolution using larger telescopes and detectors. High-speed computers have greatly enhanced the quality and speed of analysis and evaluation of these data. In this study, we are going to focus on

The Hidden Factor

the measured rotation curves of spiral galaxies and their comparison against GNM predictions.

Because of the steeper gradient in the variation of the rotation curves and contamination of bright bulge light in the central regions, it has been difficult to accurately derive the rotation curves close to the center of the galaxy. Hence, most of the earlier studies have been focused on the distribution of mass in the outer regions. Over the last few years, advanced techniques [19, 20] have been employed to determine more accurate central rotation curves in several galaxies. Center of the galaxies are still mysterious with regard to rotation properties due to a lack of resolution of the current methods at scales smaller than tens of parsecs (1 parsec is about 3 light-years). Figure 5-17 from [19] shows rotation curves of four different galaxies, which exhibit steep rotation curves near the center nucleus using high spatial and velocity resolution observations. Based on these observations, the following universal properties of the rotation curves were determined in [20] and [22]:

1. Steep central rise and peak, often starting from high velocity at the nucleus;
2. Bulge component, often causing the central peak of rotation curve;
3. Broad maximum by the disk; and
4. Halo component wherein the rotation velocity remains flat almost to the end of the luminous edge of the galaxy.

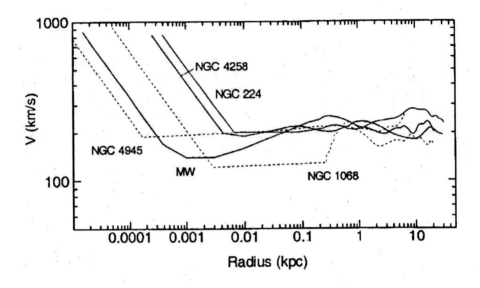

Figure 5-17: Observed high central rotational velocities in spiral galaxies. (Credit: reference [19])

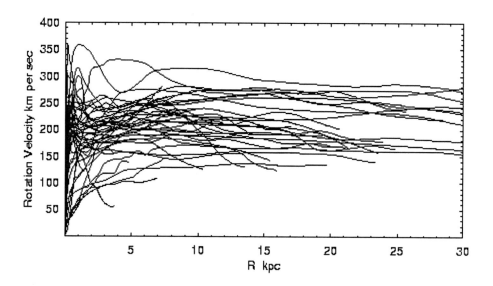

**Figure 5-18: Observed rotational velocities in spiral galaxies.
(Credit: reference [22])**

The high nuclear velocities near the center can only be detected with the highest resolution observations. Figure 5-18 from ref. [22] shows rotation curves measured for several nearby galaxies. Although these data were taken with high spatial and velocity resolution, the resolution is not sufficient to express rotation curves characteristics near the very center, particularly within 100 pc for many galaxies. For massive spiral galaxies, high nuclear rotation velocities may be universal property, however some less-massive galaxies tend to show a rigid body rise with almost zero rotation velocity at the center. It is believed that the declining rotation towards the center might be caused by insufficient angular resolution, with which position-velocity (PV) diagrams often miss the central steep rise. The true velocity may be much higher, or may start from a finite value near the center. In addition the method of deriving the rotation curve from the observed data also has an impact on the magnitude of the velocities. The study described in ref. [19] concludes that the central rotation curves derived from observed PV diagrams generally give lower limits to the actual rotation velocities.

Detailed analysis of these high rotational velocities has led several investigators [23, 24 and 25] to conclude that massive black holes exist at the center of galaxies. Consequently, the rotational velocities in the central region could be as high as the speed of light and decrease rapidly according to the equation (5-44). Using the observed rotational velocities from a sample of 26 galaxies, including 13 galaxies with newly determined black hole masses from Hubble Space Telescope measurements of stellar kinematics, the following best-fit correlation was obtained in [24], relating the black hole mass, M_{BHS} in solar masses and the velocity dispersion in km/sec:

$$M_{BHS} = 1.2 \times 10^8 \left(\frac{V_T}{200}\right)^{3.75} \quad (5\text{-}45)$$

or,

$$V_T = 200\left(\frac{M_{BHS}}{1.2 \times 10^8}\right)^{\frac{1}{3.75}} \qquad (5\text{-}46)$$

The observed flat rotation velocities in the outer disk regions of galaxies with radius R ranging from 20 to 100 kpc point to the existence of a linearly increasing mass with radius. Masses calculated using equation (5-44) range up to several times the observed range of approximately 10^{12} solar masses. In the disk of a spiral galaxy, the surface brightness is observed to decrease exponentially with increasing distance from center, while the calculated mass density from rotational velocity decreases much slower as $1/R^2$. For this reason, the variation of local mass to luminosity ratio shows a substantial increase away from the central region of the galaxy. Assuming that the Newton's laws hold in the inner and outer galactic regions, it is widely believed that massive non-luminous halos surround spiral galaxies. Although the proof of the assumption of the Newton's law is lacking, most scientists prefer this explanation to the alternative that Newtonian gravitational theory needs modification for application over galactic distances. The gravitational attraction of the non-luminous matter extending far beyond the visible limits of the galaxy leads to the flat rotation curves. There has been growing acceptance of the idea that majority of the matter that exists in the universe is dark. Understanding the physics and origin of the dark matter, both in black holes and halos of galaxies is currently a major open challenge to elementary particle physicists and astronomers in explaining the observed expansion of the universe from recent observations.

Most luminous galaxies show a slightly declining rotation curve in the outer regions beyond a flat maximum in the broad disk region. Intermediate galaxies have flat rotation curve across the disk. Less luminous galaxies exhibit rising rotational velocities with increasing distance from the center.

Avtar Singh

A Semi-empirical Understanding of galaxy rotational velocities using GNM

Equation (5-44) for rotational velocity was derived using the balance between the centrifugal force acting on a rotating body in the radial direction away from the center and the gravitational force of attraction pulling the body inward towards the center. GNM equation (5-16) has been derived using the relativistic mass-energy balance. The mechanism of luminous radiant energy (LRE) represents a direct extraction of energy from the rapidly spinning central region of the galaxy as described in reference [32]. It corresponds to an earlier theory suggesting that rotational energy could escape from the central rotating regions when it is in a strong magnetic field, which exerts a braking effect. In order to properly account for the large luminous radiant energy emanating from a galaxy, the gravitational potential energy in GNM equation (5-16) is assumed to represent the sum of luminous radiant energy (LRE) and the kinetic energy of rotation (KER), as follows.

$$\frac{3Gm^2}{5R} = LRE + KER$$

The rotational velocity observed at galactic distances of thousands of light-years from the center is much smaller than the speed of light. However, as discussed above the rotational velocity near the center may approach close to the speed of light. Hence, KER is expressed in terms of the relativistic kinetic energy as follows,

$$KER = m\,C^2 \left(\frac{1}{\sqrt{1 - \frac{V_T^2}{C^2}}} - 1 \right) \qquad (5\text{-}47)$$

and,

$$LRE = \frac{3Gm^2}{5R} - mC^2\left(\frac{1}{\sqrt{1-\frac{V_T^2}{C^2}}} - 1\right) \quad (5\text{-}48)$$

Far from the center of the galaxy (R> 100 light-years), when V_T is much smaller than C, KER is approximated by,

$$KER = \frac{1}{2}mV_T^2 \quad (5\text{-}49)$$

and,

$$LRE = \frac{3Gm^2}{5R} - \frac{1}{2}mV_T^2 \quad (5\text{-}50)$$

Simplifying the above equation, the rotational velocity in the galaxy can be determined as follows:

$$V_T = \sqrt{\frac{6Gm}{5R} - \frac{2LRE}{m}} \quad (5\text{-}51)$$

The limit of optical visibility of a galaxy is obtained when LRE approaches zero. Hence, the rotational velocity at the optical edge at the radius of R_e of the galaxy is given by,

$$V_{TRE} = \sqrt{\frac{6Gm}{5R_e}} \quad (5\text{-}52)$$

Alternatively, the limiting visibility radius R_e of a galaxy is given by,

$$R_e = \frac{6Gm}{5V_{TRE}^2} \quad (5\text{-}53)$$

Comparing equation (5-52) and (5-44), it is noted that the Newtonian mechanics and GNM provide almost similar rotational velocity near the luminous or visibility limit of the galaxy. However, in the luminous central and disk regions the rotational velocity given by equation (5-51) is smaller than that predicted by the Newtonian model due to the increasing luminosity towards the center. This results in the observed flat behavior of the rotational velocity in the galactic observations described above and further discussed later in this chapter as part of the predicted results using GNM. At distances larger than R_e, the rotational velocity is assumed to decrease as in equation (5-52):

$$V_T = \sqrt{\frac{6Gm}{5R}} \quad \text{at } R > R_e \qquad (5\text{-}54)$$

As discussed earlier, Figure 5-17 from [19] shows rotation curves of four different galaxies, which exhibit steep rotation curves near the center nucleus using high spatial and velocity resolution observations. Near to the center of a galaxy, in a region with radius $R<R_b$, the steep rise in the rotational velocity is characteristic of a large concentrated black hole mass M_{BH} for which the rotational velocity can be determined as follows:

$$V_T = \sqrt{\frac{GM_{BH}}{R}} \quad \text{at } R < R_b \qquad (5\text{-}55)$$

For the interim region, $R_b < R < R_e$, the rotational velocity (in km/sec) remains almost constant and is approximated by equation (5-46), where M_{BHS} is in solar masses:

$$V_T = 200\left(\frac{M_{BHS}}{1.2x10^8}\right)^{\frac{1}{3.75}} \quad \text{at } R_b < R < R_e \qquad (5\text{-}56)$$

Equations (5-54), (5-55) and (5-56) thus describe an empirical model for rotational velocities in all regions of galaxy from near its center to the outer edges. We would refer to this as GNM rotational velocity V_{TG} in the following sections and chapters. Equation (5-54) contains galaxy actual mass m, while equations (5-55) and (5-56) are dependent upon the black hole mass M_{bh}. From Figure 5-17, it is apparent that for the four galaxies in the database, which have almost similar rotational velocities in the outer regions, there is a large (two orders of magnitude) observed scatter in the radius R_b at which the rotational velocity levels off. The same degree of scatter also applies to the implied black hole mass M_{bh}. However, if M_{bh} is assumed to be a constant fraction of the total galaxy mass m it provides a convenient relationship to calculate rotational velocities and luminous radius R_e in terms of the galaxy mass m. As shown in Figure 5-16, the mass m of a typical galaxy remains constant and equal to the total galactic mass M_0 for radii greater than 10^{-3} kiloparsec and up to a million kiloparsec.

Figure 5-19 shows the predicted GNM rotational velocity and the observed rotational velocities of several small satellite galaxies orbiting in the gravitational field of our own Milky Way galaxy [26]. About seventy percent of all known galaxies are spiral galaxies of which Milky Way is a typical member. By using an observed rotational velocity of about 230 km/sec and an R_e of about 70 kpc from the Milky Way data, a galaxy mass of 7×10^{11} solar is obtained using equation (5-54) and a black hole mass M_{BHS} of 2.1×10^8 solar is obtained using equation (5-56). Thus a reasonable comparison with the observed data is obtained when the black hole mass is assumed to be equal to 3×10^{-4} times the total galactic mass. Using this data comparison, the following relationship is assumed to exist between the black hole mass M_{BH} and the galaxy mass m,

$$M_{BH} = 3 \times 10^{-4} m \qquad (5\text{-}57a)$$

Similarly, equating V_T from equations (5-54) and (5-55), the radius at which the rotational velocity levels off is calculated as follows,

$$R_b = \frac{5R_e M_{BH}}{6m} = 2.5 \times 10^{-4} R_e \qquad (5\text{-}57b)$$

Figure 5-20 is similar to Figure 5-19 except that the radius is plotted in a log scale to show the rising rotational velocity at smaller radii. It is to be noted that the predicted galaxy radius R_b at which the rotational velocity levels off is of the order of 10^{-2} kpc, which is in close agreement with the range of scatter of the observed data shown in Figure 5-17.

As discussed above, for a typical galaxy the mass m remains constant and equal to the total galactic mass M_o for radii greater than 10^{-3} kiloparsec. Hence, for $m = M_o$ combining equations (5-53), (5-56) and (5-57), the following approximate relationships can be obtained in terms of the leveled-off constant rotational velocity and total galaxy mass:

$$R_e = 5.09 \times 10^{-3} V_T^{1.75} \quad (R_e \text{ in kpc and } V_T \text{ in km/sec}) \qquad (5\text{-}58)$$

and,

$$R_e = 2.1 \times 10^{-4} M_0^{1.75} \quad (R_e \text{ in kpc and } M_0 \text{ in solar masses}) \qquad (5\text{-}59)$$

Figure 5-21 shows galaxy luminous radius and galaxy total mass versus the observed constant rotational velocity as predicted by GNM equations (5-58) and (5-59). As the rotational velocity increases, both the luminous radius and mass of the galaxy increases.

Figure 5-22 depicts the gravitational potential energy, luminous radiant energy and kinetic energy of rotation predicted by equations (5-15), (5-48) and (5-47) respectively for a galaxy of mass 7×10^{11} solar. At small radii, the radiant energy is approximately equal to the gravitational energy since the rotational kinetic energy is much smaller. The gravitational energy decreases as the radius increases and when it becomes of the same order as the rotational kinetic energy, the radiant energy begins to decrease rapidly until it diminishes to zero at radius equal to R_e. Beyond R_e, the gravitational and rotational kinetic energies are equal in value and decrease inversely as the radius increases according to equations (5-15) and (5-49) respectively.

Other galaxy parameters most commonly studied by scientists include surface mass density, surface luminosity density and mass to luminosity ratio. There is no direct way to measure the galactic mass distribution, and hence, it is inferred from the observed rotational dynamics of a galaxy. In most published studies, the galaxy mass distribution as a function of the radius is calculated by integrating Newtonian equation (5-44) using the measured rotational velocity. The total mass calculated in this manner includes the so-called dark matter mass, which is invisible in galactic observations and is not subject to an independent experimental or analytical validation. In the following discussion, we will describe governing equations for predicting the above parameters using GNM.

The surface mass density (SMD) at a radius R can be depicted as follows:

$$SMD = \frac{m}{4\pi R^2} \qquad (5\text{-}60)$$

The surface luminosity density is generally expressed in terms the standard luminosity of a solar mass. The equivalent mass m_L of the Luminous Radiant Energy (LRE) can be determined by expressing the LRE as an equivalent gravitational potential energy of mass m_L as follows:

$$LRE = \frac{G m_L^2}{R}$$

Simplifying the above equation in terms of m_L, the following is obtained:

$$m_L = \sqrt{\frac{(LRE)R}{G}} \qquad (5\text{-}61)$$

The surface luminosity density (SLD) can now be determined as follows:

$$SLD = \frac{m_L}{4\pi R^2} \qquad (5\text{-}62)$$

The mass to luminosity ratio (MLR) is given by,

$$MLR = \frac{m}{m_L} \qquad (5\text{-}63)$$

The absolute luminosity L is expressed in terms of the luminous radiant energy per unit time (taken by a galaxy to expand to a radius R) as follows,

$$L = \frac{LRE}{(R/C)} \qquad (5\text{-}64)$$

Figure 5-23 shows the surface mass density (SMD), equation (5-60) and surface luminosity (SLD) density, equation (5-62) of two galaxies of masses 1×10^{11} solar and 7×10^{11} solar respectively. As expected, the surface luminosity density is almost equal to the surface mass density near the center and decreases to zero at the limiting radius R_e. Radius R_e for the larger galaxy is also larger as predicted by equation (5-59). The corresponding mass to luminosity ratio (MLR) is shown in Figure 5-24.

Although, no attempt has been made in this study to make a direct comparison of the measured surface mass densities and luminosities with GNM, the overall predictions of GNM appear to have similar trends as observed in reference [22].

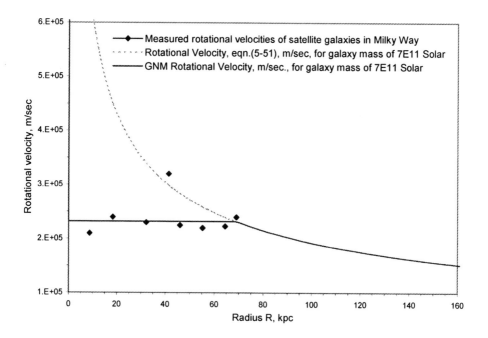

Figure 5-19: Observed versus GNM predicted rotational velocities in the Milky Way spiral galaxy.

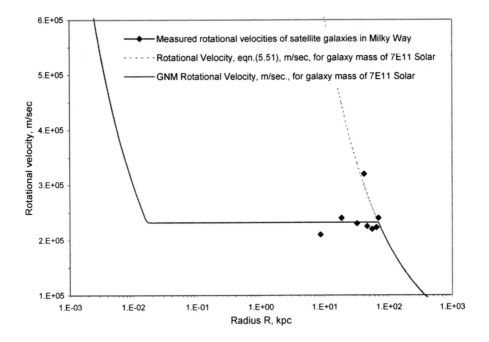

Figure 5-20: Observed versus GNM predicted rotational velocity in the Milky Way spiral galaxy

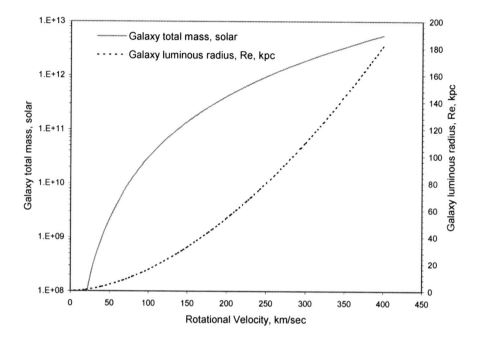

Figure 5-21: GNM predicted relationship between galaxy mass, luminous radius and rotational velocity in a spiral galaxy

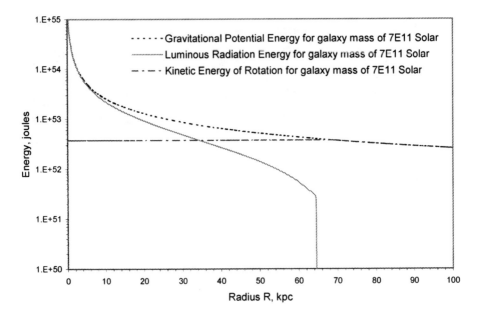

Figure 5-22: GNM predicted gravitational potential energy, kinetic energy of rotation and luminous radiation energy in a spiral galaxy.

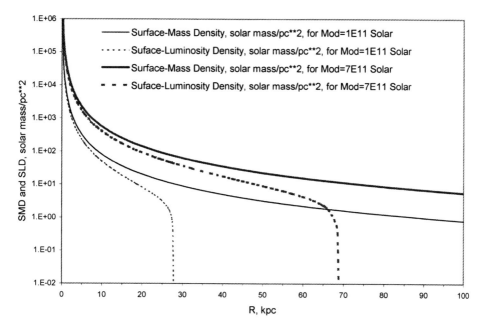

Figure 5-23: GNM predicted surface mass density and surface luminosity density in spiral galaxies.

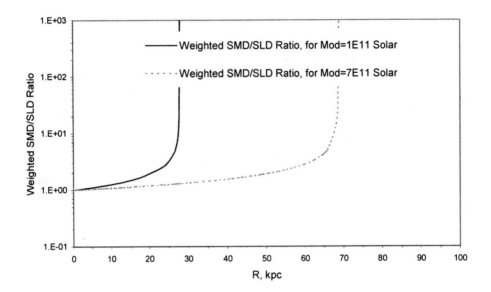

Figure 5-24: GNM predicted ratio of surface mass density to surface luminosity density in spiral galaxies

The Hidden Factor

Matter/energy ejection velocity and trajectory in a galaxy

Another prominent feature observed in the spiral galaxies is the direction of flow of luminous matter in stars contained in the galaxy. As shown in Figures 5-25 and 5-26 of two typical spiral galaxies, the radial velocity appears to decrease from the central region towards the outer regions as evidenced by the star velocity being mostly rotational at the outer visible edges. This results in an increasing angle of stars flow direction with respect to the radial direction at increasing distance from the center. For an expansion of the galaxy to occur, the matter/energy emanating from the center must have a significant radial velocity. Since high rotational velocities are also observed near the center, as discussed earlier, it is expected that the overall or resultant matter ejection or escape velocity near the center would be very high and diminish towards the outer regions due to a decreasing radial velocity. In the following discussion, we will derive analytical expressions for estimating the radial velocities using GNM and then combine them with the previously derived rotational velocities to study the overall trajectory of flow in a galaxy. Comparison of the predicted and observed trajectories will be made to validate the analytical model predictions.

The luminous radiant energy (LRE) spreads radially outward during the evolution of the galaxy as its size or radius expands. The relativistic kinetic energy corresponding to the radial velocity of expansion, V_{RE}, can be equated to LRE to obtain the following relationship,

$$LRE = m\,C^2 \left(\frac{1}{\sqrt{1 - \frac{V_{RE}^2}{C^2}}} - 1 \right) \qquad (5\text{-}65)$$

Simplifying the above equation one obtains,

$$V_{RE} = C\left[\sqrt{1-\left\{\dfrac{1}{1+\dfrac{LRE}{mC^2}}\right\}^2}\right] \quad (5\text{-}66)$$

As the galaxy expands in size, the stars have both the radial and tangential velocity components as shown in Figure 5-27. The radial component of the velocity caused by the Hubble expansion, V_{RHM}, is given by equation (5-22):

$$V_{RHM} = C\left[\sqrt{1-\left\{\dfrac{1}{1+\dfrac{H^2 R^2}{2C^2}}\right\}^2}\right] \quad (5\text{-}67)$$

The net galactic radial velocity, V_{RG}, can now be calculated as follows:

$$V_{RG} = \sqrt{V_{RE}^2 + V_{RHM}^2} \quad (5\text{-}68)$$

As described earlier, GNM rotational velocity V_{TG} is given by equations (5-54), (5-55) and (5-56) in all regions of galaxy from near its center to the outer edges. Figure 5-27 depicts a simplified schematic of the rotational, radial and total galactic velocity, V_G, which can be described as follows,

$$V_G = \sqrt{V_{RG}^2 + V_{TG}^2} \quad (5\text{-}69)$$

The angle θ of the total galactic velocity from the radial direction can now be given as follows:

$$\theta = \tan^{-1}\left(\dfrac{V_{TG}}{V_{RG}}\right) \quad (5\text{-}70)$$

The Hidden Factor

The rotational and radial velocities calculated using GNM for a galaxy with an initial mass M_o of 7×10^{11} solar are shown in Figure 5-28. The radial velocity predicted by GNM decreases from the central region towards the outer regions. This results in an increasing angle of the resultant star velocity direction with respect to the radial direction at increasing distance from the center. This is consistent with the observations shown in Figures 5-25 and 5-26 for two typical spiral galaxies, wherein the observed star velocities are mostly rotational at the outer visible edges.

Figure 5-29 depicts the same results as shown in Figure 5-28 over a much wider range and on a log scale. As discussed earlier, GNM predicts rotational and radial velocities close to that of the speed of light very near to the center of the galaxy. At increasing distance from the center, the rotational velocity decreases at a faster rate than the radial velocity until it assumes a constant value. At still farther distances, the radial velocity starts decreasing at a faster rate than the rotational velocity, until it drops to the Hubble expansion radial velocity, equation (5-22), at the limiting radius R_e. At radii larger than R_e, the Hubble expansion velocity component of the radial velocity increases until it approaches C. The corresponding variation in the angle of the resultant galactic velocity is also shown in Figure 5-29.

Figure 5.25: Whirlpool galaxy M51 or NGC 5194 (Image Credit: NASA and The Hubble Heritage Team; Acknowledgement: N. Scoville (Caltech) and T. Rector (NOAO)

Figure 5.26: Gangly spiral galaxy NGC 3184 (Image Credit: Al Kelly (JSCAS/NASA) & Arne Henden (Flagstaff/USNO))

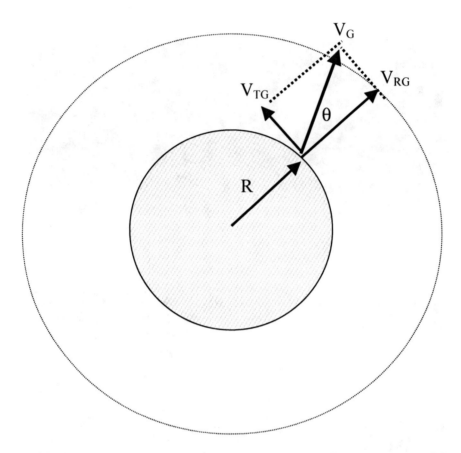

Figure 5-27: Simplified model of a galaxy expansion with radial and tangential components of velocities.

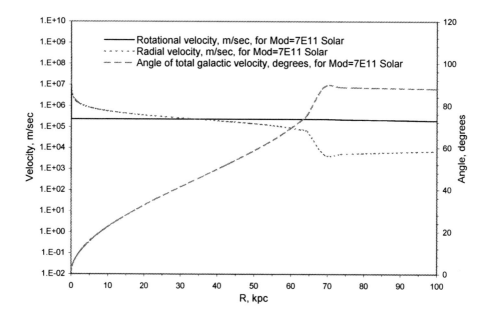

Figure 5-28: Radial and rotational velocities of stars in a galaxy predicted by GNM.

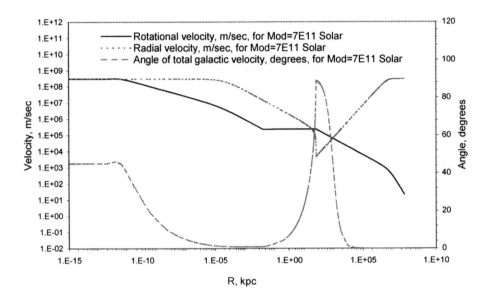

Figure 5-29: Radial and rotational velocities of stars in a galaxy predicted by GNM.

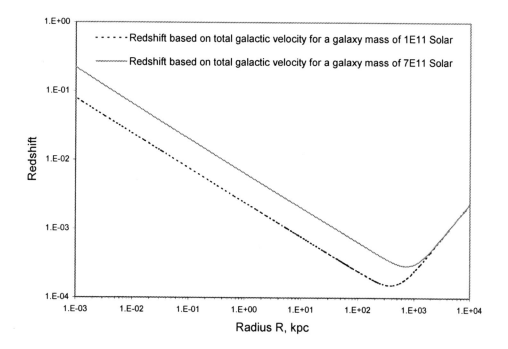

Figure 5-30: Red shift based on material ejection (total galactic) velocity in a spiral galaxy predicted by GNM.

High Red shifts Objects Related to Low Redshift Galaxies

The redshift is defined as the fractional amount by which, features such as frequency and amplitude peaks in spectra of an astronomical object are shifted to longer wavelengths or lower frequencies. The Einstein's theory of relativity [8] provides the following relationship between the redshift z and velocity V:

$$z = \sqrt{\frac{1+(V/C)}{1-(V/C)}} - 1 \qquad (5\text{-}71)$$

Using equation (5-71) and total galactic velocity, V_G, calculated red shifts are plotted in Figure 5-30 for two galaxies of masses 1×10^{11} solar and 7×10^{11} solar. The calculated redshift decreases away from the galaxy center upto a few hundred kpc before it starts increasing. It should be noted that the luminosity of the two galaxies decreases to zero at a distance less than a hundred kpc. Hence, the material ejection extends to distances far beyond the visible edges of the galaxies. As discussed earlier, this is consistent with the observations, described in reference [16], of ejection of high redshift objects, characteristic of high velocity matter ejection, in low redshift galaxies.

In summary, GNM provides a physical basis for understanding the observed data on star velocities, specifically the rotational velocities without the need for relying on dark matter in galaxies. GNM also explains potential creation of matter and corresponding objects of high redshifts observed in the vicinity of galaxies but apart from them. The gaps in physical understanding between the commonly used Newtonian models of rotational velocities and observed data is filled with the radiant energy that also explains the extent and size of the visible regions of a galaxy.

Comparison of GNM against Gödel's solution to Einstein's field equations

The BBM of the universe proceeding from nothing to an accelerated expansion is only one possibility. Gödel [27] demonstrated that a specific solution, which exactly satisfies Einstein's field equations of general relativity, corresponds to a universe rotating in a void and turning serenely like a gigantic pinwheel. In such a universe, each observer sees galaxies and all other objects in the universe rotating around him as if he were at the center of rotation. In this specific solution, galaxies are not the only things rotating. Everything in the universe including the space and time is dragged along the spirals of rotation. In addition, Gödel's solution rules out an expanding universe.

The rotational velocity equations (5-54) through (5-56) were derived from the observations of spiral galaxies. Because of the empirical fit to the observed data from galactic measurements, one would expect that these equations are valid only in the range of galactic mass and size parameters. The estimated mass and size of the universe are at least five to ten orders of magnitudes larger than those of galaxies. Hence, the extrapolation of GNM equations to the universe may be justifiable only as a first order approximation. However, the results provide astounding similarity to some of the observed characteristics of the visible universe as well as the Gödel's solution, as discussed below.

Figure 5-31 shows the absolute luminosity in watts predicted by GNM equation (5-64) for two galaxies of masses 1×10^{11} solar and 1×10^{12} solar and the universe with an estimated mass of 5×10^{22} solar. The predicted limiting radii of the luminous edges of the galaxies are of the order of 10^5 light-years and for the universe it is calculated to be 15 billion light-years. Both of these predicted limits of the galaxies and the universe are close to the observed sizes of the large spiral galaxies and the visible universe.

Figure 5-32 shows the predicted rotational, radial and total velocities and the angle of the total velocity of the universe from the radial direction. This is similar to Gödel's prediction of a rotating universe discussed above, except that GNM also predicts a significant radial velocity component, which is characteristic of an expanding universe as well. All velocities are of the order of the speed of light or 10^8 meters/sec.

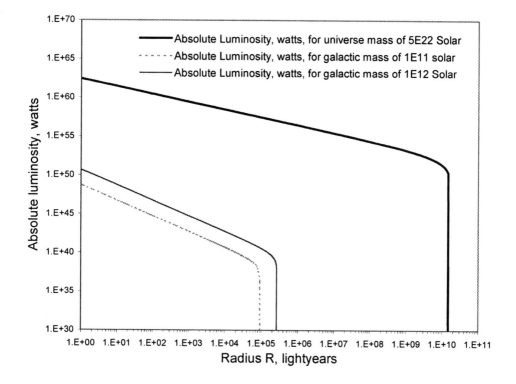

Figure 5-31: Absolute luminosity of galaxies and the universe predicted by GNM.

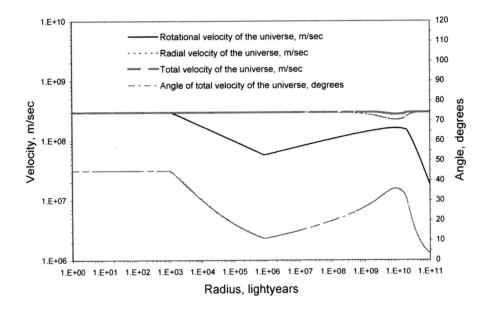

Figure 5-32: Radial, rotational and total velocities and angle of total velocity of the universe (mass of 5×10^{22} solar) predicted by GNM.

What is the role of Time? Is the universe accelerating?

As discussed earlier, based on recent observations [15] on the intensity of light from the most distant exploding star called type 1A supernova, astronomers have confirmed that the expansion of the universe is accelerating at a faster and faster rate. This is believed to be caused by an anti-gravity force or the so-called dark matter energy pushing galaxies apart. These results challenge BBM, which predicts that the pull of gravity would slow down the expansion rate of the universe as it evolves from its beginning. The new observations suggest that such a scenario predicted by BBM could have happened only during the first few billion years of the age of the universe. The later expansion was dominated by the anti-gravity energy causing the observed run-away expansion.

The GNM predicted solution of the universe is based on quasi-steady behavior, which is governed by the energy balance and is independent of considerations of time. Time is an inferred parameter simply calculated by dividing the radius of the universe by the speed of light. Hence, in the strict sense, time does not play any role in the mathematical predictions of the physical model of GNM. In GNM, as in the Einstein's theory of relativity, time is a relative entity that depends upon the speed of the observer. There is no theoretical basis at all to formulate or measure evolution history of the universe in an absolute frame of reference of time. Since there is no simultaneity in the universe due to the lack of a standard or absolute frame of reference, it makes no sense to define a point of the beginning of time (t=0), history of evolution, time rate of expansion or contraction, or an end point of time for the universe. Any definition or quantification of the history of time is merely a meaningless exercise from a universal point of view and it is difficult to assign a rational or mathematical meaning. It is conceivable to define an imaginary fixed frame of reference of time, such as the Earth's frame of reference. This could work as a reasonable approximation to describe historically the physical phenomena and events happening on earth and in its close vicinity. Such a frame of reference has no simultaneity whatsoever with the time history of events in other fast moving frame of references in the universe, such as galaxies, clusters etc.

Another common scientific assumption is that when we look into distant space away from us, we are looking into the past. The farther the object being observed, it is regarded to represent an earlier era in the age of the universe. Hence, the farthest supernovas that have been observed belong to the universe that was only a few billion years old. If this assumption were true, then an observer situated in these supernovas and looking towards us would observe us to be situated in the beginning era of the universe. In general, for different observers situated in different regions of the universe there is no unique history of the time or no correlation as to the beginning or the ending of the universe. Hence, the assumption that farther distances represent earlier times in the universe is self-inconsistent and without any physical basis. This inconsistency arises from the basic assumption that light moves from one location in space to the next in a linear fashion with a fixed speed C. As discussed in chapters 2, 3 and 4 and because of the observed non-locality in the universe, the effective speed of light itself may not be limited to a fixed value and may depend upon the speed of the observer's frame of reference. Even BBM is based on this fundamental assumption, and hence lacks consistency and realism with the observed behavior of the universe. Since time is not an absolute entity from the universe point of view, a history of the universe in any absolute time frame or for that matter its beginning (the Big Bang) and a potential ending (Crunch or death by a run-away expansion) are absurd concepts.

All states of the observable universe are quasi-steady states of being or existence that will always remain as they are, without changing with the flow of time measured on the Earth. However, the changes may be experienced to occur locally in time relative to objects of large mass having the influence of a gravitational field. At the same time, the universe consists of many a frame of references such as far away galaxies and clusters that are moving at speeds close to that of light in which the time is almost fully dilated to zero with no moving clocks. So, any objective quantification of time, which is universal, is not possible.

That also answers the question whether the universe is accelerating. It is as absurd an idea as the time itself. What is visible to the human eye and measured in the frame of reference of the earth can be deceptive with regard to the physical reality experienced by the universe.

There also have been unsettled debate and arguments as to what came first, the galaxy or the black hole? This is a classic and giant chicken-and-egg problem lurking in observable galaxies with anticipated black holes at the center. Astronomers have been tending to believe that black holes were born first and then the galaxies. Some scientists [28], however, have argued that the larger angular momentum in spiral galaxies may have kept the matter away from forming larger black holes as compared to elliptical galaxies, which are observed to contain 100 times larger black holes. This means that the black holes in spiral galaxies may have remained unchanged since their creation. Hence, the galaxies did not get formed out of the black holes, instead the opposite may have happened. This is another example of the unresolved paradoxes resulting from the assumption of the existence of an absolute time and related historical evolution of structures in the universe.

Mystery of Dark Galaxies

Existence of a large number of low mass galaxies composed entirely of dark invisible matter is hinted by both theoretical and observational investigations [29]. The theoretical basis for this is the cold dark matter theory. There are too few of the observed low mass galaxies to support the cold dark matter theory. Instead of shining like the other bright galaxies, these galaxies may be lost in the darkness of space. If such galaxies do exist, scientists believe that the existing theories of structure formation may need to be revised or replaced. Recently, astronomers have reported [30] the discovery of a cluster of galaxies via its weak gravitational lensing effect on background galaxies.

In the conventional BBM, galaxies are believed to have formed from large volumes of gas produced in the big bang. Due to fluctuations in density of these gas clouds, the gas in the denser patches would have pulled in the surrounding material by the attractive force of gravity. But the formulation of structures controlled by the gravitational pull of the visible gas alone would be a much slower process than what is actually observed in the universe and the galaxies would still be forming even today. This observed faster rate of galaxy formation is believed to be due to the existence of invisible dark matter that may outweigh the normal visible matter in the universe by orders of magnitudes.

Some scientists [31] believe that there is something fundamentally wrong with the existing BBM theories about the structure formation in the early universe. Cosmologists rely on the assumption of small density variations caused by quantum fluctuations in the first fraction of a second following the big bang, for early structures to form. These small fluctuations are assumed to magnify by the inflationary expansion of the universe. The usual assumption is that the universe was as lumpy on small scale as on the large scale. Scientists believe that this assumption does not support the observation of shortage of small galaxies and small clusters.

How does GNM explain this mystery of dark galaxies? As shown in Fig. 5-21, the limiting visible radius of a galaxy approaches to very small values for a galaxy mass less than 1×10^8 solar. For a galaxy with a mass of the order of 1000 solar masses, GNM would predict a visible radius of approximately 15 light-years. Such galaxies located at large distances from the earth would be hardly detectable even with powerful telescopes. The key here is that GNM predicts the existence of dark or low luminosity galaxies without the theory of dark matter, which is unable to provide a consistent explanation of the formation and existence of the observed structures in the universe. Another key point supported by GNM is that the observed mass of structures in the universe can be satisfactorily predicted without resorting to their historical evolution with respect to time. Time, again, is shown to be an ineffective and non-physical parameter in explaining the observed behavior of the universe with regard to the visible as well as dark galaxies.

Chapter 6

Unraveling of Quantum Mechanics Using the Gravity Nullification Model

Quantum mechanics theory, in spite of its successes, has remained enigmatic and paradoxical because of a lack of understanding of its inner workings. Scientists are still puzzled over its features such as wave-particle duality, entanglement, "spooky action at a distance" and whether or not God plays dice with the universe. The behavior of the small versus the classical large objects shows uncommon features that still remain unexplained by existing theories. Common human experience does not support the intuitions of quantum mechanics. Freeman Dyson, an eminent physicist, expressed this strangeness by saying- "I understand now that there isn't anything to be understood." Even Richard Feyman, known for his mastery of the subject, raised the question: "How does it work? What is the machinery behind the law...We have no ideas about a more basic mechanism from which these results can be deduced."

In Chapter 3, we investigated the relativistic mass-energy behavior and how it effects not only space-time, but also the fundamental fabric of the universe in terms of non-locality and action at a distance. We derived the Gravity Nullification Model to explain the observed non-locality in the universe. In this and the following Chapters we will explore the inner workings of quantum mechanics using GNM and address the following remaining issues in the famous debate between Albert Einstein and Neils Bohr:

♦ Is there an inherent and irreducible uncertainty in prescribing both the position and momentum of quantum entities or particles (the Heisenberg Uncertainty)?

- Is wave-particle behavior a duality or complimentarity? What are the limits of wave versus particle behavior? What is behind quantum behavior of small particles?
- Is quantum mechanics a complete theory and does it represent reality? Why does it work for small particles and not for classical objects in the real world?

Heisenberg's Uncertainty

The first question above involving uncertainty is related to the observed dual behavior of photons and other small particles such as electron, proton etc. in the microscopic world that act both as particles as well as waves. The phenomenon of wave-particle duality was described in detail in Chapter 4. In the macroscopic world, the effects of this duality behavior are not noticeable for objects, which are physically orders of magnitudes larger than their effective wavelengths. The wavelength or the actual physical size of an object, whichever is greater, describes the region in space over which the object spans its existence. For quantum particles, the wavelength can be a lot larger than the physical size of the particle leading to an uncertainty of defining the exact location and momentum simultaneously. In quantum mechanics, the wavelength relates to the probability of finding the particle in that region. Larger the wavelength associated with a particle, larger is the probability of finding that particle and vice versa. However, there is an uncertainty associated with specifying the exact location and velocity (or momentum) of the particle in the region of the wavelength. This is known as the Heisenberg Uncertainty. This uncertainty is often presumed to occur due to the direct and unavoidable impact of the measuring device or process on the motion or spatial location of the measured entity itself, especially when the measured entity is a small microscopic particle. Light is often used to observe small particles. When a photon of light strikes a particle being observed, a fraction or all of the momentum of the photon may be transferred to the particle impacting its velocity and position in an unpredictable manner. This unpredictability gives rise to the uncertainty, discovered by Heisenberg in 1927, in specifying the current or future position and momentum of the particle.

As discussed in Chapter 4, in GNM this uncertainty directly depends upon the relative physical size of the particle and the size of the wavelength, which in turn depend upon its mass and velocity. However, in classical physics and quantum mechanics, mass, space and time are considered to be fixed (Newtonian frame of reference) and independent entities, which are used to define velocity and momentum of the particle. In classical quantum mechanics, the wavelength of the particle is defined as follows [1],

$$\lambda_{dbr} = \frac{h}{mV} \qquad (6\text{-}1)$$

Where λ_{dbr} is de Broglie wavelength, h is Planck's constant, m is the mass and V is the velocity of the particle in the Newtonian frame of reference. This wavelength describes the amount of uncertainty in the position of the particle.

$$\Delta x \approx \lambda_{dbr} \qquad (6\text{-}2)$$

In order to detect the particle, the wavelength of the photon and its momentum must be of the order of magnitudes of wavelength and momentum of the particle. Of course, the uncertainty in the measurement could be much higher, since the relative sizes and momentums of the photon could be much larger than that of the particle, which are unknown and multiple number of photons may be required to make the measurement. The product of the uncertainty in spatial position x and momentum mV can be expressed as below,

$$(\Delta x)\{\Delta(mV)\} \geq h \qquad (6\text{-}3)$$

Detailed calculations by Heisenberg determined the following mathematical form of his uncertainty principle,

$$(\Delta x)\{\Delta(mV)\} \geq \frac{h}{2\pi} \qquad (6\text{-}4)$$

Since Planck's constant, h, is very small, the uncertainties at macroscopic level or for scales of everyday large objects are negligible. However, at microscopic level wherein we deal with small particles such as electrons, protons and atoms, the uncertainty becomes significant. The fundamental probabilistic formulation of quantum mechanics has its roots in the Heisenberg's uncertainty.

Evaluating Heisenberg's Uncertainty using GNM

Before we perform a quantitative evaluation of the Heisenberg's uncertainty using GNM, let us discuss some of the fundamental assumptions underlying the principle. As discussed above, the following inherent assumptions are built into the Heisenberg's uncertainty principle:

1. Fixed space and time: Both the spatial position and velocity measurements are presumed to be made by an observer in a Newtonian inertial frame of reference having fixed space and time.
2. Non-zero velocity or momentum: The uncertainty principle holds only for a non-zero velocity or momentum. Hence, in an inertial frame of reference moving with the object itself wherein the velocity or momentum is zero, the principle is violated.
3. Wave-particle behavior is assumed to be governing the state of the object being observed. Conditions wherein such a behavior may be suppressed or superceded by the classical macroscopic scale experience are ignored. Neither any criterion that defines the range of applicability of the wave-particle behavior is addressed by the Heisenberg's uncertainty principle.

It is apparent from items 1 and 2 above that the Heisenberg's principle is an extension of the Newtonian mechanics and should be valid only for macroscopic objects moving at speeds much lower than the speed of light. For particles with higher speeds approaching the speed of light, such as electrons and photons, the relativistic effects become significant and the assumption of fixed space and time does not hold. Heisenberg's principle does not specify the impact of relativistic effects on the uncertainty. Ironically, the principle provides the fundamental basis for probabilistic formulation of the quantum behavior of the fast moving quantum particles, which are expected to have strong relativistic effects.

The Gravity Nullification Model (GNM), which is based on the specific theory of relativity, supports the notion that the wave-particle behavior is a fully deterministic and not a probabilistic or uncertain phenomenon. From GNM perspective, the root cause of the uncertainty is the very assumption of fixed space and time as well as ignorance of the relativistic effects, which remain unaccounted for in the Newtonian based approach adopted by Heisenberg and the quantum theory. In other words, GNM would demonstrate that the uncertainty of the Heisenberg principle is not inherent in nature or ontological, but rather a direct result of choosing an inappropriate frame of reference (the Newtonian fixed space and time) to describe a physical event or phenomenon. When the relativistic effects and their effect on space/time are properly accounted for in observing the particle motion, the uncertainty would diminish to zero without any lower limits. In the following, we will use GNM to revaluate the Heisenberg's principle and calculate physical parameters and conditions under which the principle may hold true and vice versa.

In Chapter 4, equations (4-6) and (4-8) were derived for calculating the self-decaying mass and wavelength as a function of velocity. The equations are given below:

$$m = M_o \sqrt{1 - (V/C)^2} \qquad (4\text{-}6)$$

$$\lambda_{sdm} = \frac{hV}{M_o C^2 \sqrt{1-(V/C)^2}} \qquad (4\text{-}8)$$

Using a velocity derivative of equation (4-6), the uncertainty in momentum mV can be calculated as follows,

$$\Delta(mV) = M_o \left(\frac{1 - 2\left(\frac{V}{C}\right)^2}{\sqrt{1-\left(\frac{V}{C}\right)^2}} \right)(\Delta V) \qquad (6\text{-}5)$$

Substituting equations (4-8) and (6-5) into equation (6-4) and simplifying, the following is obtained for a self-decaying mass,

$$(\Delta x)\{\Delta(mV)\} \geq \frac{h}{2\pi}\left[2\pi\left(\frac{V}{C}\right)\left\{ 1 - \frac{\left(\frac{V}{C}\right)^2}{1-\left(\frac{V}{C}\right)^2} \right\} \Delta\left(\frac{V}{C}\right) \right] \qquad (6\text{-}6)$$

Let us now define a Heisenberg's Uncertainty Factor, HUF_{sdm}, for a self-decaying mass as follows:

$$HUF_{sdm} = \left[2\pi\left(\frac{V}{C}\right)\left\{ 1 - \frac{\left(\frac{V}{C}\right)^2}{1-\left(\frac{V}{C}\right)^2} \right\} \Delta\left(\frac{V}{C}\right) \right] \qquad (6\text{-}7)$$

Equation (6-6) then can be rewritten as follows:

$$(\Delta x)\{\Delta(mV)\} \geq \frac{h}{2\pi}(HUF_{sdm}) \qquad (6\text{-}8)$$

Comparing the above against the Heisenberg's principle, equation (6-4), the principle is satisfied when the absolute value of HUF≥1 and violated when the absolute value of HUF<1. It is apparent from equation (6-7) that HUF depends directly upon the value of velocity V and its uncertainty ΔV. If the uncertainty in velocity, ΔV, is zero, HUF will be zero and hence, the Heisenberg's principle will be violated. HUF given by equation (6-7) for a self-decaying mass for non-zero values of ΔV is shown in Figure 6-1. It is to be noted that for velocities less than approximately 70% of the speed of light C, the HUF is less than 1 and the Heisenberg's principle is violated even for an uncertainty ΔV being as high as 90%. For smaller uncertainties in V, the principle is satisfied only for much larger velocities. For 1% uncertainty in V, the minimum value of V is greater than 95% of the speed of light C for the principle to hold true.

Using a similar approach as above, the Heisenberg's Uncertainty Factor, HUF_{ndm}, for a non-decaying mass can be derived using equations (2-3), (4-12) and (6-6) as follows:

$$(\Delta x)\{\Delta(mV)\} \geq \frac{h}{2\pi}(HUF_{ndm}) \qquad (6\text{-}9)$$

wherein,

$$HUF_{ndm} = \left[2\pi \left(\frac{V}{C}\right) \left\{ \frac{1}{\sqrt{1-\left(\frac{V}{C}\right)^2}} \right\} \Delta\left(\frac{V}{C}\right) \right] \qquad (6\text{-}10)$$

Comparing the above against the Heisenberg's principle, equation (6-4), the principle is satisfied when the absolute value of HUF≥1 and violated when the absolute value of HUF<1. As for the self-decaying mass above, HUF for non-decaying mass depends directly upon the value of velocity V and its uncertainty

The Hidden Factor

ΔV. If the uncertainty ΔV is zero, HUF will be zero and hence, the Heisenberg's principle will be violated. For non-zero values of ΔV, HUF for a non-decaying mass is shown in Figure 6-2. For velocities less than approximately 40% of the speed of light C, the HUF is less than 1 and the Heisenberg's principle is violated even for an uncertainty in velocity V being as high as 90%. For smaller uncertainties in V, the principle is satisfied only for much larger velocities. For 1% uncertainty in V, the minimum value of V is greater than 99% of the speed of light C for the principle to hold true.

In summary, GNM provides the physics behind the Heisenberg's Uncertainty principle, its limitations and boundaries within which it holds true.

- GNM results provide a physical understanding of why objects in the microscopic and macroscopic world moving at much lower velocities than the speed of light, do not exhibit uncertainties predicted by the principle.
- GNM accounts for the fundamental weakness in the formulation of the Heisenberg's principle, which is the ignorance of the relativistic effects and relying on the Newtonian mechanics for describing the motion of small particles in motion. It is well known that Newtonian laws, which provide a satisfactory representation of physical reality at small velocities, are inappropriate and incomplete to describe high-speed relativistic phenomena that govern the motion of quantum or small particles.
- GNM reveals that there is no inherent uncertainty in nature insofar as the true relativistic nature of the reality is realized and treated as such. The Heisenberg's uncertainty results entirely from a common mindset that all reality can be described in terms of fixed space and time or a non-relativistic frame of reference. Under the disguise of the new-age quantum mechanics, Newtonian laws still pervade the modern scientific method.
- Since the Heisenberg's principle forms the fundamental basis for quantum mechanics, science needs to reconsider Einstein's remark questioning the completeness of the quantum mechanics theory.

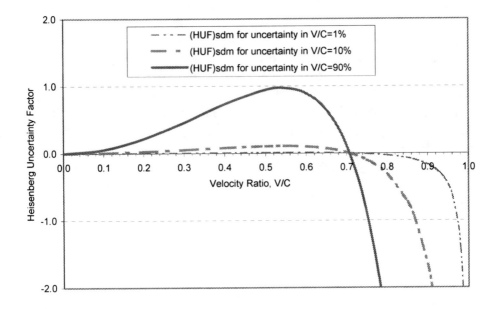

Figure 6-1: Heisenberg's Uncertainty Factor for a self-decaying mass.

The Hidden Factor

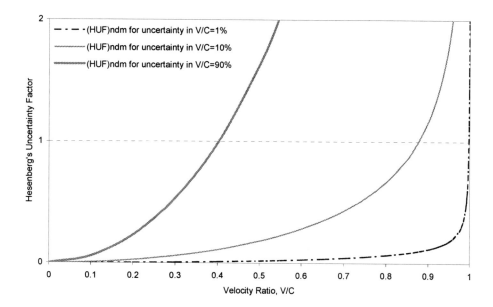

Figure 6-2: Heisenberg's Uncertainty Factor for a non-decaying mass.

Avtar Singh

'What is a Quantum Particle?' - Physical Limits of Quantum Behavior

A quantum particle is an illusive entity that can appear from or disappear into nothingness or vacuum, and exhibits unexplained behavior that follows weird rules involving strange properties. The quantum behavior or properties are so far different from those of the real life objects that there appears to exist separate worlds or universes for the ordinary real life objects versus the quantum objects. In fact, the theory of parallel universes is one of the highly regarded theories explaining the quantum weirdness. There still exists a big gap in the fundamental understanding of the duality that exists between the behaviors of the small microscopic quantum particles versus the behavior of classical macroscopic objects. In the discussion below, we will develop a GNM based understanding of the quantum behavior and derive mathematical expressions of the physical limits that govern transition between quantum and classical behavior.

In Chapter 5, the following equation (5-16) representing GNM based universe model was obtained from conservation of relativistic mass-energy, kinetic energy and gravitational potential energy,

$$(M_0 - m)C^2 = mC^2 \left\{ \frac{1}{\sqrt{1-\left(\frac{V}{C}\right)^2}} - 1 \right\} + \frac{3Gm^2}{5R} \qquad (6\text{-}11)$$

As discussed in chapter 4, the fundamental characteristic of quantum behavior is the wave-particle complimentarity, which allows a quantum entity to act as a wave or particle depending upon its environment. When the particle can move uninhibited, it generally acts or exists as a wave, and when it is intercepted via a measuring device or a fixed boundary wall it appears to act as a particle. Such spontaneity of converting from a particle (mass) to wave (energy) is built into the observed wave-particle behavior in a variety of experiments involving quantum particles such as electrons, atoms or molecules. In Chapter 3, we described spontaneous motion of a body wherein a conversion of mass to kinetic energy is allowed according to the equation (3-5) given below,

$$(M_0 - m)C^2 = mC^2 \left(\frac{1}{\sqrt{1 - \frac{V^2}{C^2}}} - 1 \right) \quad (3\text{-}5)$$

After simplifying, the above equation can also be written in the following form:

$$m = M_o \sqrt{1 - (V/C)^2} \quad (6\text{-}12)$$

Comparing equation (3-5) above and GNM equation (6-11), it is evident that the spontaneous decay exhibiting the wave-particle behavior can occur when the gravitational potential energy is much smaller and negligible compared to the kinetic energy. Based on this, the following criterion can be obtained for a quantum entity to exhibit a spontaneous and uninhibited wave-particle behavior:

$$\frac{3Gm^2}{5R} \leq mC^2 \left\{ \frac{1}{\sqrt{1 - \left(\frac{V}{C}\right)^2}} - 1 \right\} \quad (6\text{-}13)$$

Radius R in the above equation can be approximated by the wavelength of the self-decaying mass given by equation (4-8),

$$R \approx \lambda_{sdm} = \frac{hV}{M_o C^2 \sqrt{1-(V/C)^2}} \qquad (4\text{-}8)$$

Substituting equations (6-12) and (4-8) into equation (6-13) and simplifying leads to the following:

$$M_{oq} \leq \left(\sqrt{\frac{hC}{G}}\right) \sqrt{\frac{5V}{3C} \left\{ \frac{1-\sqrt{1-\left(\frac{V}{C}\right)^2}}{\left(1-\left(\frac{V}{C}\right)^2\right)^{\frac{3}{2}}} \right\}} \qquad (6\text{-}14)$$

Wherein, M_{oq} represents the maximum or limiting self-decaying rest mass that exhibits the wave-particle duality or quantum behavior. Masses greater M_{oq} would be expected to behave as classical rather than quantum. It should be noted that Planck's mass is defined as,

$$M_{pl} = \sqrt{\frac{hC}{G}} \qquad (6\text{-}15)$$

Using equation (6-14), we now define Planck's Mass Factor (PMF) as follows:

$$PMF = \sqrt{\frac{5V}{3C} \left\{ \frac{1-\sqrt{1-\left(\frac{V}{C}\right)^2}}{\left(1-\left(\frac{V}{C}\right)^2\right)^{\frac{3}{2}}} \right\}} \qquad (6\text{-}16)$$

Combining equations (6-15) and (6-16) gives the following criterion for a mass, expressed in terms of a non-dimensional mass M_{oqn}, to act as a quantum particle:

$$M_{oqn} = \frac{M_{oq}}{M_{pl}} \leq (PMF) \qquad (6\text{-}17)$$

Similarly, combining equations (6-14) and (4-8) and simplifying, the following can be obtained:

$$\lambda_q \geq \left(\sqrt{\frac{hG}{C^3}}\right)\left[\frac{\left(\frac{3V}{5C}\right)}{\left\{\frac{1}{\sqrt{1-\left(\frac{V}{C}\right)^2}} - 1\right\}}\right] \qquad (6\text{-}18)$$

Wherein, λ_q represents the minimum or limiting wavelength of a self-decaying mass that exhibits the wave-particle duality or quantum behavior. It should be noted that Planck's length is defined as,

$$L_{pl} = \sqrt{\frac{hG}{C^3}} \qquad (6\text{-}19)$$

Using equation (6-18), we now define Planck's Length Factor (PLF) as follows:

$$PLF = \left[\frac{\left(\frac{3V}{5C}\right)}{\left[\left\{\frac{1}{\sqrt{1-\left(\frac{V}{C}\right)^2}}-1\right\}\right]} \right] \quad (6\text{-}20)$$

Combining equations (6-19) and (6-20) gives the following criterion for the critical wavelength, expressed in terms of a non-dimensional wavelength λ_n, to act as a quantum particle:

$$\lambda_n = \frac{\lambda_{sdm}}{L_{pl}} \geq (PLF) \quad (6\text{-}21)$$

Using similar approach for a non-decaying mass and combining equations (4-12) for the wavelength of a non-decaying mass along with equation (6-13), the following governing equations can be developed for the limiting (maximum) non-decaying rest mass, M_{oq}' and limiting (minimum) wavelength, λ_q', Planck's Mass Factor, PMF', and Planck's Length Factor, PLF', which exhibit the wave-particle duality or quantum behavior:

$$M_{oq}' \leq \left(\sqrt{\frac{hC}{G}}\right)\left[\sqrt{\frac{5V}{3C}}\left\{\sqrt{1-\left(\frac{V}{C}\right)^2} - \left[1-\left(\frac{V}{C}\right)^2\right]\right\}\right] \quad (6\text{-}22)$$

$$\lambda_q' \geq \left(\sqrt{\frac{hG}{C^3}}\right) \frac{\left(\frac{3V}{5C}\right)}{\left[\left\{\frac{1}{\sqrt{1-\left(\frac{V}{C}\right)^2}}-1\right\}\right]} \quad (6\text{-}23)$$

$$PMF' = \left[\sqrt{\frac{5V}{3C}}\left\{\sqrt{1-\left(\frac{V}{C}\right)^2}-\left[1-\left(\frac{V}{C}\right)^2\right]\right\}\right] \quad (6\text{-}24)$$

$$PLF' = \frac{\left(\frac{3V}{5C}\right)}{\left[\left\{\frac{1}{\sqrt{1-\left(\frac{V}{C}\right)^2}}-1\right\}\right]} \quad (6\text{-}25)$$

Figure 6-3 shows the critical quantum rest mass and wavelength for a self-decaying mass predicted by GNM equations (6-14) and (6-18) respectively. For most velocities the predicted critical or maximum quantum rest mass is approximately of the same order as Planck's mass of 2.1767×10^{-8} kilograms. For very small velocities, the critical rest mass decreases and for velocities close to the speed of light C, it increases by several orders of magnitude. Hence, at small velocities the quantum behavior is experienced only by very light particles and at higher velocities even a lot heavier particle can exhibit quantum properties or wave particle duality. For example, Figure 6-3 shows that an entity of a Planck's mass will act as a non-quantum or classical object at velocities less than half the speed of light; at higher velocities the same object will act as a quantum or particle-wave. Thus GNM explains the behavior of certain particles that may behave both as classical as well as quantum objects depending upon their speed of motion. A very light particle of a mass several orders of magnitude lower than Planck's mass will always behave as a quantum particle even at small velocities. Figure 6-3 shows that the predicted critical or minimum wavelength is close to Planck's length of 1.616×10^{-35} meters for most velocities. At very low velocities, the critical wavelength increases and for higher velocities approaching the value of C, it decreases significantly.

Figure 6-4 shows the critical quantum rest mass and wavelength for a non-decaying mass predicted by GNM equations (6-22) and (6-23) respectively. The key difference from the self-decaying mass behavior occurs at high velocities close to the speed of light wherein the critical quantum mass decreases for the non-decaying mass. As a result, the maximum non-decaying mass that can exhibit the quantum behavior is always less than Planck's mass. A non-decaying mass that is heavier than Planck's mass, will always act as a non-quantum or classical object. The critical wavelength behavior for non-decaying mass is similar to the self-decaying mass.

Figure 6-5 and 6-6 depict the same information shown in Figures 6-3 and 6-4 in a non-dimensional form represented by non-dimensional factors PMF, PLF, PMF' and PLF' given by equations (6-16), (6-20), (6-24) and (6-25) for a self-decaying and non-decaying mass respectively. For masses greater than the critical quantum mass, the region for clasical or non-quantum behavior extends up to velocities approximately 70% of the speed of light for a self-decaying mass and close to the speed of light for a non-decaying mass. Since enormous external energy input is required to accelerate a non-decaying mass to the speed of light, practically all non-decaying masses will exhibit non-quantum or classical behavior. A self-decaying mass spontaneously transforms to convert to kinetic energy, hence is self-sufficient to attain large velocities without needing external energy input. Thus a spontaneous self-decaying mass can exhibit quantum or wave particle behavior irrespective of its large rest mass.

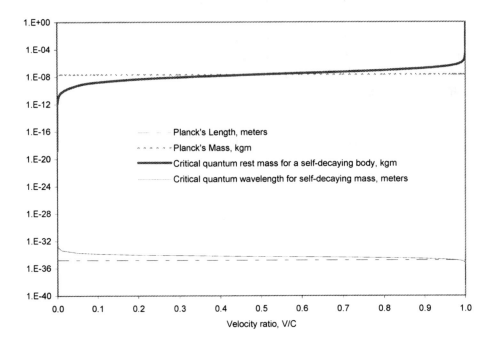

Figure 6-3: Critical quantum rest mass and wavelength for a self-decaying mass.

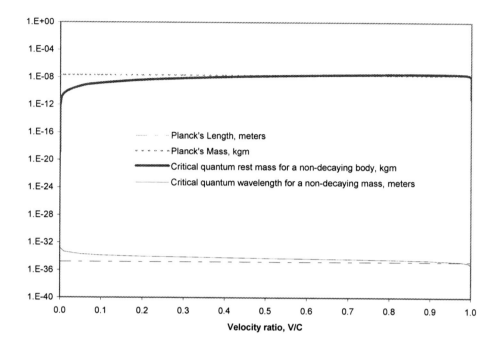

Figure 6-4: Critical quantum rest mass and wavelength for a non-decaying mass.

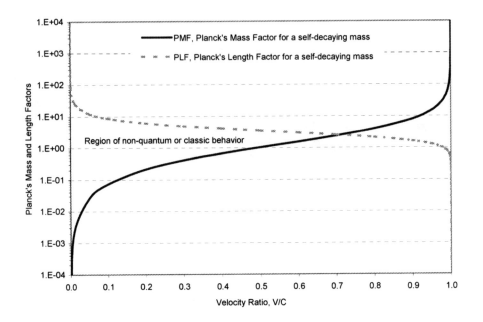

Figure 6-5: Planck's Mass and Length Factors for a self-decaying mass.

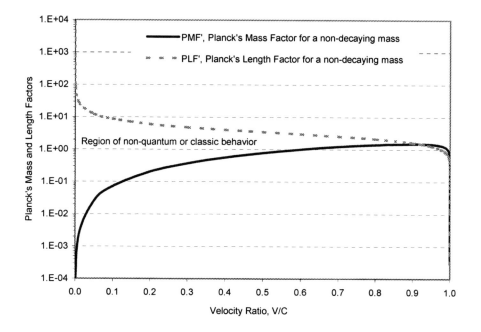

Figure 6-6: Planck's Mass and Length Factors for a non-decaying mass.

Avtar Singh

Quantum Paradoxes

Now we will discuss how the above GNM based mathematical and physical understanding can explain some of the weird and mysterious aspects of quantum mechanics. Roger Penrose [33] states-

"I mentioned the rather disturbing fact that, in our description of fundamental physics, we use two quite different ways of describing the world, depending upon whether we are talking about the large-scale or the small-scale end of things.

...I believe that the normal view of physicists is that, if we really understood quantum physics properly, we could deduce classical physics from it. I want to argue differently. In practice, one does not do that – one uses *either* the classical level *or* the quantum level."

At the quantum level, the state of a system is described by a complex-number weighted superposition of all possible alternatives. The time-evolution of the quantum state or the Schrödinger evolution is obtained by the linear superposition of all possible states. Each of the individual states are assumed to evolve independently, but are superposed together with complex number weightings that are invariant in time. This linearity is built into the widely accepted formulation of quantum mechanics in Schrödinger's wave equation, which represents a deterministic and quantitative description of quantum reality. However, as Roger Penrose [33] points out, the rules change when a measuring device or observer looks into the quantum reality and in the process coverts it into a classical reality. This process of conversion from quantum to classical reality is defined as the *Collapse of the Wave Function* or *Reduction of the State Vector*. An example of this is the observation of an electron wave as a dot when it hits the screen. Following this process, the two alternatives are no longer superposed linearly. Instead, the squares of the weighting complex numbers become the ratio of the probabilities of the two alternatives. The reality that was fully deterministic before the collapse, all of a sudden becomes non-deterministic or probabilistic after the collapse of the wave function caused by the process of measurement or observation by a conscious observer. This procedure itself is regarded by all professional scientists as a fundamental theory of quantum mechanics.

This paradox or mystery of quantum mechanics theory is also known as the *Measurement Problem*. The most famous example of this is the so-called *Schrödinger's Cat* paradox in which a cat can exist in a state of being both dead and alive at the same time. A cat, which is a classical object, is never seen to exist in such a quantum state in real life experience.

Another famous mystery of quantum mechanics is the *Quantum Entanglement* or *Quantum Non-locality*. Einstein and his colleagues, Podolsky and Rosen, first highlighted the physical problem involved in this paradox. Subsequent experiments performed by Alan Aspect [2] have confirmed the correctness of the predictions of quantum mechanics. In Aspect's experiments photon pairs are emitted at a source in an entangled state and travel in two different directions to two detectors located at about 12 meters apart from each other. The decision to measure the direction of polarization of the two separated photons was made after the photon were in full flight from the source to the detectors. The results of the measurements showed that the two photons do not behave as two separate and independent classical objects. Instead, the observed states of the two photons were observed to be in the entangled state that matched the predicted joint probabilities of quantum mechanics. Similar experiments have been performed to verify quantum entanglement or non-locality over distances of several kilometers. Highlighting this unexplained mystery, Roger Penrose [33] states-

"I should emphasize that, in these *non-local* effects, events occur at separated points A and B, but they are connected in mysterious ways. The way in which they are connected – or entangled – is a very subtle thing. They are entangled in such a way that there is no way of using that entanglement to send a signal from A to B – this is very important for the consistency of quantum theory with relativity. Otherwise, it would have been possible to use quantum entanglement to send messages faster than light. Quantum entanglement is a very strange type of thing. It is somewhere between objects being separate and being in communication with each other – it is a purely quantum mechanical phenomenon and there is no analogue of this in classical physics."

GNM Solution to the Quantum Measurement Problem

Roger Penrose [33] then goes on to offer the following hints and exploratory suggestions to handle the quantum measurement problem-

"What we do not have is a thing which I call OR standing for *Objective Reduction*. It is an objective thing – either one thing *or* the other happens objectively. It is a missing theory."

"The viewpoint I am advocating is that something goes wrong with the superposition principle when it applies to significantly different space-time geometries... Somehow this superposition actually becomes one OR the other and it is at the level of space-time that this happens."

"What is the relevance of Planck's length, 10^{-33} cm, to quantum state reduction? ...When the (space-time) geometries start to differ by that amount you have to worry what to do and it is then that the rules might change...when there is sufficient mass movement between the two states under superposition such that the two resulting space-time differ by something of the order of 10^{-33} cm."

"I am going to regard the superposition of the one state plus the other as an unstable state – it is a bit like a decaying particle or a uranium nucleus or something like that, where it might decay into one thing or another and there is a certain time-scale associated with that decay. It is a hypothesis that it is unstable, but this instability is to be an implication of the physics we do not understand.

....The thing is that, in the superposed state, you have to take into account the gravitational contribution to the energy in the superposition. But you cannot really make local sense of the energy due to (quantum) gravity and so there is a basic uncertainty in the gravitational energy...That is just the sort of thing which one gets with unstable particles."

Now, we will discuss how the mathematical formulations of GNM account for the phenomena identified above by Roger Penrose in explaining the physical basis behind the quantum measurement problem. First of all, GNM equation (6-11) used above to derive the criteria for quantum behavior of particles, accounts for the contribution of the gravitational energy as suggested by Roger Penrose. Hence, the uncertainty caused by the so-called quantum gravity is eliminated in GNM. Secondly, GNM equation (6-11) properly treats the energy conservation involving spontaneous decay of a mass. In addition to the gravitational potential energy, the kinetic energy and the mass energy are properly accounted for in GNM. This provides for a proper consideration of the mass movement or conversion between various states that a particle can experience before, during or following a measurement is made. As the mass decays, its kinetic energy or velocity increases with a corresponding dilation of space/time or non-locality as discussed earlier in Chapters 2, 3 and 4. GNM thus inherently accounts for the mass-energy movement and superposition of different space-times or mass-energy states associated with different possible states of the quantum particle.

In summary, GNM includes the missing physics described by Roger Penrose in dealing with the mysterious paradoxes of the quantum mechanics. Let us now see how GNM can explain what happens physically during the process of measurement on a quantum particle and how a quantum particle changes to a classical entity or how the *Objective Reduction* occurs.

Figure 6-7 shows a schematic of this process. Let us assume that there exists a quantum entity with a rest mass equal to about one tenth of Planck's mass and moving at approximately half the speed of light (V/C = 0.5) in free space as depicted by the right tail end of the arrow in Figure 6-7. When this quantum entity (existing dominantly as a wave in the free space is interrupted by a classical measuring device, its velocity practically stops to zero (V/C = 0), as depicted by the left end of the arrow in Figure 6-7. As is evident, the process of measurement causes a sudden change or the so-called *collapse* of the quantum wave (also described as a wave function) into the region of non-quantum or classical behavior. The Objective Reduction occurs because of this sudden change in the quantum wave velocity-ratio to zero leading to the classic behavior as a particle. Note that if the initial mass and velocity of the quantum entity and the final velocity imparted by the measuring device are exactly known, then the final state of the classical particle will be fully deterministic. However, if the initial state parameters are not fully known, then the final classical state will not be deterministic, but will have an associated uncertainty that will depend upon the uncertainty inherent in the initial state parameters. GNM thus explains the physical process involved in the *Objective Reduction* or

Collapse of the Wave Function that occurs during the *Measurement problem* in quantum mechanics.

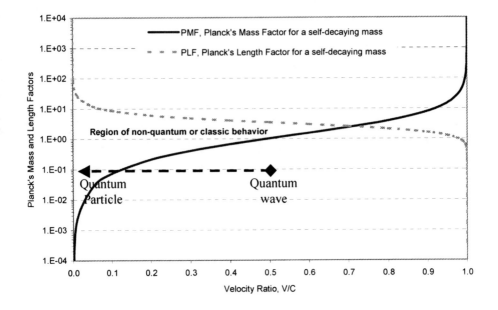

Figure 6-7: Objective Reduction of quantum to classical behavior during the process of measurement using a classical device.

GNM Solution to the Quantum Entanglement or Quantum Non-locality

Now let us discuss how the mystery of the *Quantum Entanglement* or *Quantum Non-locality* is explained by GNM in a physical and mechanistic manner. In chapter 2, we demonstrated that the non-locality results from dilation of space-time when the speed V of an entity or its frame of reference equals C. In chapter 3, we provided a more generic basis, which is the conservation of mass-energy for non-locality in the universe. In chapter 4, we described how GNM predicts non-locality of a self-decaying mass that acquires an infinite wavelength when its velocity approaches C.

In Chapter 2, it was shown that due to the dilation of time in a frame of reference moving at close to the speed of light, the effective velocity of light for an observer situated in the stationary frame of reference appears to be much larger than C. The arguments discussed in Chapter 2 are repeated here to address the non-locality paradox. Figure 2-3 shows the variation of effective velocity, $V_{eff,}$ versus V measured in the stationary frame of reference. When V equals zero, V_{eff} equals V_o as expected in the stationary frame of reference. When V approaches C, V_{eff} approaches infinity. This means that if an object were moving at speed of light in the stationary frame of reference, its effective speed in its own frame of reference would become infinity. This is shown in Figure 2-4, wherein C_{eff} represents the effective speed of light in a frame of reference moving at velocity V relative to the fixed frame of reference and C_0 represents speed of light on the fixed frame of reference. As seen in Figure 2-4, C_{eff} remains approximately equal to C until V/C approaches 0.8. As V approaches C, C_{eff} becomes much larger than C approaching infinity at V equal to C. This provides a possible explanation of action at a distance or connectivity in the universe predicted by quantum mechanics and observed in various experiments involving experiments with light.

The non-locality was also discussed in chapter 3, wherein it was shown that the non-locality is mandated in nature by conservation of mass-energy in all possible frames of references. Since the total mass-energy in all frames of references must be conserved, no frame of reference is independent of any other frame of reference moving with a different speed. It implies that any two frames of references moving relative to each other must transfer mass-energy between them to allow conservation of the total mass-energy. Mass-energy dilation in one frame of reference, as given by equation (3-2), is transferred to the other frame to cause equal and opposite mass-energy expansion such that the total mass-energy of the universe remains constant. Such transfer of mass-energy is possible only through seamless connectivity or non-locality in the universe.

In the Newtonian Mechanics that represents our understanding of everyday experience, an isolated object or mass can be accelerated to acquire different velocities depending upon the amount of force applied. The object moving at a uniform speed equal to any of the possible values of the velocity represents a distinct inertial frame of reference which is assumed to be independent of other frames of references. What is ignored, however, is the fact that the applied force comes from the neighboring frame external to the frame of reference of the moving body and the source of this force is the mass-energy in the neighboring or the connected frame. The connected frame could either be the stationary frame or any other frame moving at a different velocity than the body itself. Because of the absence or lack of consideration for this exchange or connectivity, mass-energy is not conserved in different Newtonian frames of references. At velocities much smaller than C, the relativistic effects being small, this error in the non-conserved mass-energy is small compared to the total mass-energy of the universe. This leads to the common everyday experience or the delusion of the isolated non-locality of the moving body from the rest of the universe.

To clarify this aspect of the Newtonian inertial frames of references, let us consider again the two frames of references shown in Figure 2-1. Let us now consider a non-moving body in the stationary (V=0) frame of reference. The same body when viewed from the moving frame of reference has a speed that is equal and opposite to the speed V of the moving frame. Hence, the mass-energy of the body in the moving frame is higher than the mass-energy in the stationary frame by the amount of the kinetic energy acquired by the body. Based on the Newtonian laws of motion, this increase in kinetic energy is caused by the action of an isolated force on the body in motion. Such an understanding of the motion observed in everyday experience is so deeply rooted in us that it has become a matter of habit for us to assume non-locality a foregone reality in the universe. Such an understanding has no scientific basis since it violates the fundamental law of conservation of mass-energy on a universal basis.

Now, let us consider a mass that decays spontaneously according to the Gravity Nullification Model, equation (3-2), and attains a speed V. Because of the increased speed, the space and time dilate as per equations (2-1) and (2-2). If the mass completely decays to zero, V becomes equal to C and space and time both dilate to zero. As discussed in Chapter 2, the complete dilation of space and time leads to the observed non-locality and action-at-distance.

It is argued and implied in Einstein's Specific Theory of Relativity that a real signal that carries information in the form of an energy wave with a finite non-zero frequency and wavelength can never travel faster than C. This is consistent with the modified Postulate 3 described in Chapter 2 or the Gravity Nullification Model as follows. The information in a signal is stored in the form of energy waves of specific frequency and wavelength as seen by an observer in the stationary frame of reference. Since the signal maintains a finite energy or non-zero rest mass that constitutes the specific stored information during its transmission through empty space, its speed V remains slightly less than (however very close to the value of) C and the space-time does not completely dilate to zero. This preserves the locality of the signal since its velocity V in the stationary frame is limited by C. A photon of light, on the other hand, that is not restricted to carry a signal of a non-zero rest mass, can spontaneously decay its mass to zero as it travels through the empty space. Hence, non-locality and coherence is observed in the behavior of a photon over the entire universe, while a signal carrying a significant and measurable mass-energy exhibits a local behavior with its propagation velocity limited by C.

In chapter 4, we described how GNM predicts non-locality of a photon based on a self-decaying mass behavior that can have an infinite wavelength at velocities close to C. The spatial extent of the wavelength determines the distance over which the entanglement or non-locality of a particle is expected to occur. The correlation or coherence over this spatial extent is instantaneous and not subject to the locality caused by the finite speed of light, which limits the speed of travel of a signal. Figure 4-14 provides a physical picture of the photon behavior as a self-decaying mass during the processes of its emission or absorption at a stationary surface. At the surface, its velocity is zero to match the stationary boundary condition and its particle rest mass is equal to the equivalent quantum energy of the photon. During its emission, the mass of a photon spontaneously decays to provide kinetic energy for its motion away from the surface. The velocity of the photon accelerates as the mass converts to kinetic energy until it decays completely to attain the speed of light C. The self-decaying characteristic of the photon mass negates the need for any external force or source of energy to accelerate the photon to the speed of light and provides a physically consistent explanation of the photon wave-particle behavior. During the absorption process of an incoming photon at a stationary surface, the above sequence of events is reversed wherein the velocity of the photon decreases from C to zero and the mass increases from zero to the rest mass.

Frequencies and wavelengths of a self-decaying photon with a rest mass of M_0, can be calculated using equations (4-7) and (4-8) for varying values of velocity. Figure 4-12 shows photon wavelengths predicted by equations (4-8) for two different photon masses of 1×10^{-45} kilogram and 1×10^{-35} kilogram respectively. At zero velocity, the wavelength predicted by GNM equation (4-8) is zero and as V increases to C the wavelength increases to infinity. Since the wavelength is inversely proportional to the rest mass, the wavelength for the lighter rest mass is proportionally longer than the heavier one. In order to demonstrate the non-locality of a very light mass particle due to its dramatically large wavelength, we will calculate the ratio of the wavelength of a self-decaying mass given by equation (4-8) to the radius or size of the universe predicted by GNM Equation (5-21) describing the Relativistic Hubble Model (RHM):

$$\lambda_{sdm} = \frac{hV}{M_o C^2 \sqrt{1-(V/C)^2}} \qquad (4\text{-}8)$$

$$R = \sqrt{\frac{2C^2}{H^2}\left\{\frac{1}{\sqrt{1-\left(\frac{V}{C}\right)^2}} - 1\right\}} \qquad (5\text{-}21)$$

Dividing the above two equations and simplifying gives the following,

$$\frac{\lambda_{sdm}}{R} = \frac{1}{\sqrt{2}} \frac{hH}{M_o C^2} \left[\frac{\frac{V}{C}}{\sqrt{\sqrt{1-\left(\frac{V}{C}\right)^2} - \left\{1-\left(\frac{V}{C}\right)^2\right\}}} \right] \qquad (6\text{-}26)$$

For small velocity V<<C, the above equation simplifies to the following,

$$\frac{\lambda_{sdm}}{R} = \frac{hH}{M_o C^2} \qquad (6\text{-}27)$$

We can now calculate the rest mass of a particle, defined here as the God-particle for which the wavelength is equal to the Hubble radius of the universe as follows,

$$M_o = \frac{hH}{C^2} \qquad (6\text{-}28)$$

For a Hubble Constant of 2.27E-18 sec.$^{-1}$, the rest mass of a God-particle is calculated to be 1.66×10^{-68} kilograms. Figure 6-8 shows the ratios of wavelengths to the Hubble radius calculated by equation (6-26) for a Planck's mass, proton, electron and the God-particle. As is apparent, the ratios of wavelength to Hubble radius for a Planck's mass, proton and electron are several orders of magnitude smaller ($< 10^{-38}$) than the ratio of 1 for the God-particle. Hence, the God-particle has equal and 100% probability to be at all points in the universe, satisfying an ideal non-locality, as compared to the protons or electrons, which will be located or confined to much smaller regions of their wavelengths. Also, the God-particle will appear to be in perfect coherence and entanglement all over the universe because of its wavelength encompassing the entire universe.

In summary, GNM provides a mathematical and physical understanding of the observed *Quantum Entanglement* or *Quantum Non-locality*.

GNM Solution to the Paradox of Particle Spin

Why does a particle have a spin? GNM provides a physical model based on gravitational potential energy of the particle as described in chapter 5. Figure 6-9 shows the rotational velocities as a function of radius predicted by GNM equations (5-54), (5-55) and (5-56). When the size or radius of the particle is very small, the rotational velocity could be as high as the speed of light. For larger sizes, the rotational velocity decreases to smaller values. When a quantum measurement is made on a particle by a classical measurement device, the size or region to which the particle is confined is very small and hence leading to a substantial rotational velocity.

GNM thus provides a mathematical and physical based model to the particle spin as well as the wave behavior of a particle observed in quantum mechanics.

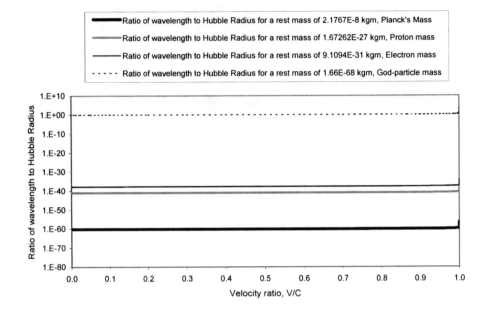

Figure 6-8: Ratio of wavelength to Hubble Radius for varying particle rest mass.

The Hidden Factor

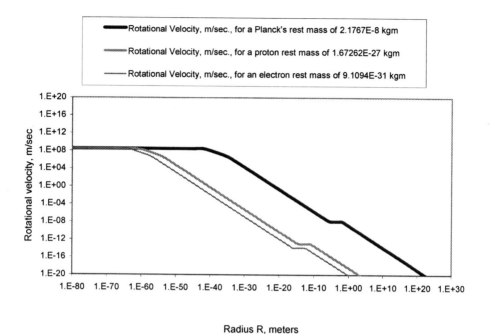

Figure 6-9: Rotational velocity predicted by GNM for an electron, proton and Planck's mass.

GNM Solution to the Quantum Entanglement in the Bose – Einstein Condensate

When some liquids or ideal gases are cooled to sufficiently low temperatures (below 2.2 degree Kelvin) they exhibit a strange behavior called superfluidity. A superfluid exhibits no viscosity as a fraction of the atoms of the gas undergo a process called Bose-Einstein Condensation (BEC) [4]. As an ideal gas or liquid is cooled down, the motion of its individual atoms slows down and due to an increase in density the atoms begin to pack up closer together. Then, at very low temperatures close to the absolute zero degree Kelvin, the slowest moving atoms are seen to grow thousands of times from their original size overlapping each other and completely loosing their well-defined boundaries. These wraithlike atoms can pass through each other and flow against gravity. The Bose-Einstein condensate may consist of hundred of atoms which have turned into a mysterious fluid defying all classical laws of fluid mechanics and loosing its classical fluid properties. All atoms completely loose their individual identity in a so-called quantum entanglement with one another.

The criterion for BEC to occur is that the average distance between the particles of the gas must be smaller than the wavelength (from wave-particle duality principle). When this occurs, the state of the two atoms overlap causing them to behave as if they are in the same mode also called "uniform quantum phase". BEC is a thermo-dynamical process that occurs in very nearly ideal gases which do not transition easily to either conventional liquid or solid phase when cooled to extremely low temperatures. The following are some of the key conditions for BEC to occur:

1. Detailed statistical analysis can be performed to show that when particles behave like *identical entities that are indistinguishable from each other*, only then BEC phase transition can occur. Such particles are called *Bosons*.

2. Atoms are cold and densely packed moving at extremely low velocities relative to each other. This allows them to be within a distance of effective wavelength from each other causing *phase coherence* or *quantum entanglement* to occur. When this occurs, the particles exhibit a unified mode or behavior without any relativity among them.

The above conditions can be achieved when the particles or atoms are confined without a direct physical contact with a classical room-temperature apparatus. The condensate traps are generally very small in size, typically several micrometers in diameter. The confinement is achieved via a magnetic trap wherein reasonably high densities of atoms at extremely low temperatures can be attained by firing suitably designed laser beams. During the last few years, BEC has been achieved with rubidium, lithium, sodium and hydrogen atoms.

Another key observation that is important for understanding the BEC phenomenon is that the measured speed of light in typical BEC is very small. Scientists [34, 35] experimenting with BECs have measured speed of light close to almost zero to just a few meters per second in cold vapor of sodium ions.

We will now show how GNM predicts the onset of BEC based on the wavelength of Lithium atoms moving at small velocities and with a speed of light of a few meters per second as observed for a typical BEC. First of all, for a particle to behave as a wave, GNM provides a self-decaying mass model wherein the mass can completely convert to a waveform of energy. The wavelength for a self-decaying mass is given by equation (4-8):

$$\lambda_{sdm} = \frac{hV}{M_o C^2 \sqrt{1-(V/C)^2}} \qquad (4-8)$$

Figure 6-10 shows the predicted wavelengths of a Lithium atom for a range of velocity ratios (V/C) close to 1 and speed of light values in the range of a fraction to several meters per second. GNM predicts that for the expected range of these velocities in a typical BEC, the wavelength of a Lithium atom is within the range of the observed size, several micrometers, of the BECs. This allows the individual atoms to be within a distance of effective wavelength from each other resulting into a *phase coherence* or *quantum entanglement*, and leading to the formation of a BEC.

As discussed above, another required feature of the BEC via statistics is that BEC phase transition can occur only when particles behave like *identical entities that are indistinguishable from each other*, such as Bosons. GNM provides a physical basis for this apparent statistical property. The particles can act identically and indistinguishably only when they act completely as waves and not as particles confined in space and time. The self-decaying mass defined by GNM is exactly such a physical entity that can transition spontaneously to an energy-wave when it is allowed to move without a classical constraint, as is the case with magnetically trapped BECs. This could also explain why a BEC can not be formed when confined by rigid walls of a classical container. The rigid or non-moving walls of a container inhibit free motion of the particles resulting in a zero velocity and thus inhibiting the wave behavior of the particles.

In summary, GNM provides a physical based understanding of the BEC phase transition and quantum entanglement.

The Hidden Factor

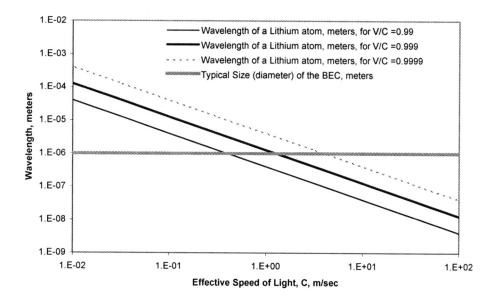

Figure 6-10: Wavelength of a Lithium atom in a Bose-Einstein condensate as a function of the effective speed of light.

Avtar Singh

Theory of Parallel Universes Explained by GNM

The theory of parallel universes [2] has been advanced by its proponents to explain the mysterious collapse of the wave function consisting of infinite number of states to one classical solution experienced by the observer. Each of the infinite number of universes corresponds to a probable classical outcome that can occur when an observer looks at the quantum system. The parallel universes coexist but are not quite parallel in the strict sense, since they do communicate with each other at the quantum level through a common space and time. An observer can not experience them all and can detect them indirectly via their impact on the observed events in space-time.

The proponents argue that the theory of parallel universes is by far the simplest in explaining the observed quantum experiments since it involves the fewest additional assumptions. When an observation is made, it is presumed that the whole ensemble of parallel universes partitions in two groups with different outcomes. The observer happens to experience only one of the outcomes in his own universe. The other dilemma solved by this theory is the so-called observer paradox in quantum cosmology. In a classical quantum observation, the observer is outside the quantum system and looking at it causing the collapse of the wave function. However, when the observer is within or part of the quantum system itself, such as in the case of observations of the universe by an observer within it, the standard interpretation of the quantum theory fails. Hence, the quantum cosmological observations can not be explained by the standard interpretation of quantum theory. The theory of parallel universes eliminates this dilemma. A third point forwarded as an advantage of this theory is that it does not require the strict definition as to who the observer is. The definition of an observer relative to the quantum system is an open question in the standard interpretation, and probably involves a definition of the consciousness, which is not an easy problem to resolve.

The theory of parallel universes eliminates the observer and mind from interpretation of the observed reality. The key weakness of the theory is that it can not be tested and involves many open questions with regard to the properties and nature of mass-energy-space-time interactions among many universes.

Now, let us discuss how GNM provides a physical understanding of the many possible outcomes signifying parallel universes, of an observation depending upon the characteristics of the observer. According to GNM, the mass-energy-space-time dilation is directly dependent upon the relative velocity ratio (V/C) between the observer and the observed. Two different observers traveling at different speeds experience different mass-energy-space-time and hence different physical realities. Figure 6-11 shows GNM predicted mass-space-time dilation as a function of the velocity of the observer as well as the progressions of manifold outcomes of the parallel universes. At low velocities, the mass-space-time act as fixed or classical as assumed in the Newtonian mechanics. Different events and outcomes can exist for different observers moving at different speeds relative to each other at lower velocities. However, at large velocities when V approaches C, the mass-space-time dilate to almost zero and hence the outcomes or events/phenomenon observed by different observers experience quantum entanglement becoming completely coherent with each other. In a way, higher velocity represents a higher level of the observer consciousness leading to a non-locality or non-relativity in what is observed. The theory of many parallel universes reflects many a less than perfectly conscious observer (V<<C) leading to many possible outcomes.

In summary, GNM provides a qualitative but physical understanding of the theory of parallel universes based on mass-space-time dilation in a unified single universe and subject to its one set of laws. The many universes phenomenon is correlated via different mass-space-time characteristics of different observers correlated via their speed or consciousness.

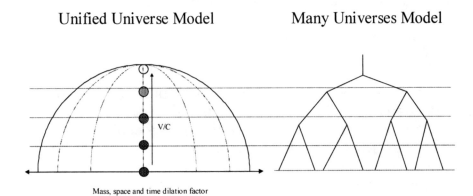

Figure 6-11: Depiction of GNM single universe model versus parallel many universes model of quantum mechanics.

Chapter 7

Evaluation of Quantum and Classical Effects of Gravity using the Gravity Nullification Model

In Chapter 6, we investigated the classical versus quantum behavior predicted by GNM that includes the effects of gravity and relativistic kinetic energy. Although gravitation was one of the first and fundamental phenomenon that received a thorough mathematical treatment by Newton in early developments of physics, it continues to remain a significant challenge to modern scientists to integrate gravity with other observed forces of nature in a consistent and seamless mathematical model. Especially, a viable and common mathematical description of gravity in quantum and relativity theories has eluded scientists for the last several decades with no convergence or success in sight. All attempts to quantize gravity in existing quantum theories run into serious mathematical problems. Paul Davies [4] states-

"So long as gravity remains an unquantised force there exists a devastating inconsistency at the heart of physics. Although quantum effects of gravitation are unlikely to have any detectable results in particle physics (and precious few elsewhere, save possibly in early-universe cosmology), nevertheless it is vital that a consistent quantum description be found, otherwise gravitation cannot be properly interfaced with the rest of physics."

One of the major predictions of the general relativity theory, which is the most fundamental and successful theory of gravitation thus far, is the phenomenon of gravitational collapse wherein the matter, without an anti-gravity force, collapses under its self-attraction forces of gravitation to a point leading to a singularity of infinite density. The general theory of relativity is unable to predict what if any special phenomenon may happen below Planck's scale that may eliminate the singularity of gravitational collapse. In this chapter, we will focus on how GNM eliminates this singularity and provides a consistent mathematical description of the gravitational phenomenon at both large and extremely small scales to even smaller than Planck's size, 10^{-35} meters.

The quantum behavior or properties of particles are so far different from those of the real life objects and things that there appears to exist separate worlds or universes for the ordinary real life objects and the quantum objects. However, there still exists a big gap in the fundamental understanding of the gravitational effects that govern the behavior of small microscopic quantum particles versus the behavior of classical macroscopic objects. In the discussion below, we will develop a GNM based understanding of the effects of gravity at the quantum versus classical scales and derive mathematical expressions of the physical limits that govern gravitational stability of quantum and classic objects. Specifically, we will investigate possible explanations to the following questions:

- What role does gravity play in exhibiting the stability of a mass or particle as a function of its size?
- What role does gravity play in governing the relationship between Planck's mass and Planck's length?
- What role does gravity play in governing the behaviors exhibited by Bose-Einstein condensates at low temperatures?

In Chapter 5, the following equation (5-16) representing GNM based universe model was obtained from conservation of relativistic mass-energy, kinetic energy and gravitational potential energy,

$$(M_0 - m)C^2 = mC^2 \left\{ \frac{1}{\sqrt{1-\left(\frac{V}{C}\right)^2}} - 1 \right\} + \frac{3Gm^2}{5R} \quad (5\text{-}16)$$

Figure 7-1 shows the predicted ratio of actual mass m to total mass M_o as a function of radius R predicted by GNM equation above for an electron, proton and Planck's mass. The velocity ratio in the above equation is calculated by the Relativistic Hubble Model equation (5-22),

$$\frac{V}{C} = \sqrt{1 - \left\{\frac{1}{1 + \frac{H^2 R^2}{2C^2}}\right\}^2} \quad (5\text{-}22)$$

As shown in Figure 7-1, Planck's mass becomes unstable below a radius of the order of 10^{-35} meters, which is the same order of magnitude as Planck's length calculated by the following relationship:

$$L_{pl} = \sqrt{\frac{hG}{C^3}} \quad (7\text{-}1)$$

Thus GNM predicts that for a mass greater than Planck's mass to remain gravitationally stable as a classical mass, its radius must be greater than Planck's length. Only those masses that are smaller than the Planck's mass can exist as classical or gravitationally stable masses at a size smaller than Planck's length. GNM thus provides a physical basis for Planck's Mass and Planck's Length as the limits of gravitational stability for a classical mass.

Similarly, for a proton and electron, the gravitational stability radii are predicted to be of the order of 10^{-54} meters and 10^{-58} meters respectively, which are about 19 to 23 orders of magnitude smaller than Planck's length.

It is to be noted that for very small values of R shown in Figure 7-1, velocity ratio V/C is much smaller than 1, and hence, the kinetic energy term in equation (6-11) above is much smaller than the gravitational energy term. Neglecting the kinetic energy term leads to the following simplified GNM equation for gravitational stability at sizes below Planck's scale applicable to quantum particles:

$$(M_0 - m)C^2 = \frac{3Gm^2}{5R} \qquad (7\text{-}2)$$

Simplifying this one obtains the following expression for the critical Gravitational Stability Radius (GSR):

$$R_{cg} = \frac{3}{5} \frac{Gm^2}{(M_0 - m)C^2} \qquad (7\text{-}3)$$

At the initiation of gravitational instability, m is of the same order (but smaller than) as the rest mass M_0. Hence, the above equation can be approximated and simplified to provide an order of magnitude approximation for the critical gravitational stability radius as follows:

$$R_{cg} \approx \frac{GM_0}{C^2} \qquad (7\text{-}4)$$

Figure 7-2 shows the Gravity Stability Radius predicted by GNM equation (7-4) over a wide range of masses from less than the electron mass to about 100 solar masses. For a solar mass (2×10^{30} kgm), the predicted Gravitational Stability Radius is of the order of a kilometer. If a solar mass is confined to a radius less than a kilometer, it will violently convert to energy, almost in a way of an explosion (just like a supernova or Big Bang), due to the excessive gravitational energy or force. Figure 7-3 shows this gravitational instability in terms of the ratio of the actual to the total mass as predicted by GNM equation (7-3) for a solar mass.

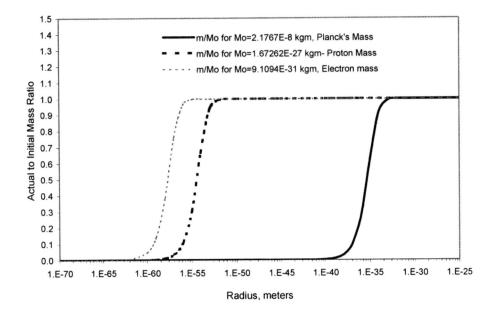

Figure 7-1: Gravitational stability radius for an electron, proton and Planck's mass predicted by GNM.

The Hidden Factor

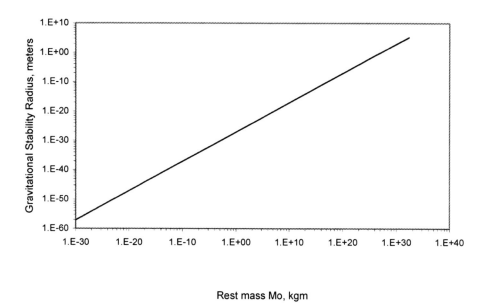

Rest mass Mo, kgm

Figure 7-2: Gravitational stability radius as a function of mass predicted by GNM.

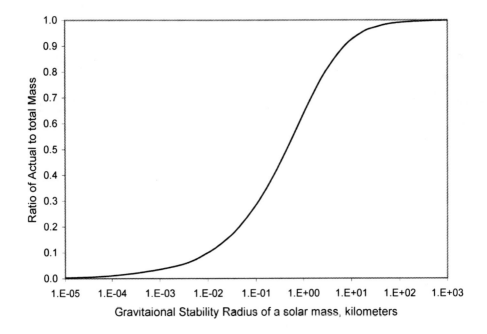

Figure 7-3: Gravitational stability radius of solar mass predicted by GNM.

Derivation of Planck's Length and Planck's Mass using GNM

Using equation (7-4) and Planck's law, it can be shown that the critical gravitational stability radius predicted by GNM is equivalent to Planck's length as follows. Planck's law for a photon of frequency f and energy E is given by:

$$E = hf \qquad (7\text{-}5)$$

Since, E is equal to the total mass-energy $M_o C^2$, the above can written as:

$$M_0 = \frac{hf}{C^2} \qquad (7\text{-}6)$$

Substituting the above in equation (7-4),

$$R_{cg} \approx \frac{Ghf}{C^4} \qquad (7\text{-}7)$$

Now, for a photon the critical gravitational stability radius can be approximated by its wavelength below which it will act as a classical mass,

$$\lambda = \frac{C}{f} = R_{cg} \qquad (7\text{-}8)$$

Substituting in equation (7-7) and simplifying using equation (7-1), one gets:

$$R_{cg} \approx \sqrt{\frac{hG}{C^3}} = L_{pl} \qquad (7\text{-}9)$$

Combining equations (7-6), (7-8) and (7-9), the following can be obtained for Planck's mass:

Avtar Singh

$$M_{pl} = \sqrt{\frac{hC}{G}} \qquad (7\text{-}10)$$

Once again, it has been shown that GNM provides a physical basis for relationship between gravitational stability radius, Planck's mass and Planck's length.

Anti-gravity versus Gravity

No physical description of matter and gravity is complete without the anti-gravity. If there is no anti-gravity, matter cannot exist because it would be annihilated by the gravitational pull inwards. So for a gravitationally stable mass to exist, the gravitational force or energy must be balanced by the anti-gravitational force or energy. The Cosmological Constant, Λ, described in Chapter 5 represents the anti-gravity energy of the vacuum. This anti-gravity energy is responsible for the observed expansion of the universe. The anti-gravity energy that leads to a gravitationally stable mass is defined as the Cosmological Anti-gravity parameter, Λ_{ag} as follows. In Chapter 5, the Cosmological Constant was defined by equation (5-19):

$$\Lambda = \frac{6}{R^2}\left\{\left(\frac{M_0}{m}-1\right)-\frac{3Gm}{5RC^2}\right\} \qquad (5\text{-}19)$$

Let us define the Cosmological Anti-gravity parameter by the following equation:

$$\Lambda_{ag} = \frac{6}{R^2}\left(\frac{M_0}{m}-1\right) \qquad (7\text{-}11)$$

Substituting in equation (5-19),

$$\Lambda = \Lambda_{ag} - \frac{6}{R^2}\left(\frac{3Gm}{5RC^2}\right) \qquad (7\text{-}12)$$

or,

$$\Lambda_{ag} = \Lambda + \left(\frac{18Gm}{5R^3C^2}\right) \qquad (7\text{-}13)$$

In Chapter 5, the Cosmological Constant was described by the equation (5-23),

$$\Lambda = \frac{3H^2}{C^2} \qquad (5\text{-}23)$$

As discussed in Chapter 5 under the section describing the Cosmological Constant problem, so far no credible physical theory has been advanced that could predict even the right order of magnitude of the vacuum energy responsible for the observed acceleration of the universe. Particle-physics theories put forward possible partial explanation for existence of vacuum energy and a non-zero Cosmological Constant, but the values predicted by these are about 120 orders of magnitude greater than the recent supernova observations. If the vacuum energy of such magnitude were existent, the existing particle theories and the Big Bang theory would predict an acceleration of almost infinite magnitude that would rip apart atoms, stars and galaxies. Clearly, the current understanding of the vacuum energy is incorrect.

Figure 7-4 shows the value of the Anti-gravitational Cosmological parameter, Λ_{ag}, calculated by equation (7-13) for masses of an electron and Planck's mass. For large radii, the value of Λ_{ag} is equal to the Cosmological Constant Λ. As radius decreases, the value of Λ_{ag} increases, by several orders of magnitude, due to an increase in the gravitational potential energy. However, the rate of increase of Λ_{ag} decreases when the radius decreases below the critical gravitational stability radius due to a corresponding decrease in the mass that converts to energy. Hence, the Anti-gravitational Cosmological parameter for a non-zero or finite mass is several orders of magnitude larger at small radii than the Cosmological Constant for the vacuum. In fact, the value of Λ_{ag}, calculated by equation (7-13) for Planck's mass at a radius equal to Planck's length is of the order of 10^{70} m^{-2}, which is about 120 orders of magnitude larger than the vacuum Cosmological Constant Λ (~10^{-52} m^{-2}), matching the predictions of the particle physics models.

GNM thus provides a physical and mechanistic model that predicts the Cosmological Constant for the vacuum as well as the Cosmological Anti-gravity Parameter for a mass as a function of particle mass and radius. The missing physics from the existing particle physics and relativistic theories is represented by the anti-gravity phenomenon that impacts the quantum or classical behaviors at small and large scales respectively.

The Hidden Factor

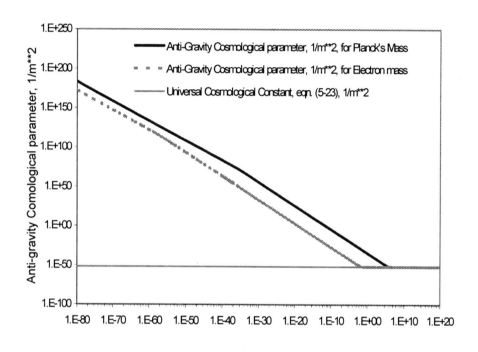

Figure 7-4: Anti-gravitational parameter for an electron and Planck's mass.

Avtar Singh

Resolving Black Hole Controversies

Most scientist believe, based on predictions of current theories, that a massive star can implode due to its gravitational instability to form an extraordinary dense object called the black hole. There is no observed evidence of an actual black hole other than indirect observations of its impact on other objects or matter in the vicinity of it. Recent research [36] performed by two physicists, Emil Mottola of the Los Alamos National Laboratory in New Mexico and Powel Mazur of the University of South Carolina in Columbia, have pointed to some "humiliating" problems with existing theories of black holes. Some of these problems are discussed below:

1. Since in the existing theories there is no force opposing gravity, scientists believe that a massive star at the end of its life would explode in a powerful supernova explosion. At the end of this explosion, if a core of twice the solar mass remains it would be gravitationally unstable leading to the formation of a black hole singularity with infinite mass density where all the laws of physics would break down.
2. The current theories of black holes predict an enormous, billions and billions times higher entropy than that of the forming star. The question is that what is the credible source of such enormous entropy.
3. The infinite density of the black hole would lead to an event horizon of infinite space-time curvature. The gravitational pull would swallow any photons approaching the black hole event horizon, and would gain infinite amount of energy by the time they reach the event horizon. Where does such an enormous energy come from?

Because of the above contradictions and inconsistencies, it is believed that black holes may not exist in reality after all and could be the result of incomplete understanding of the physics of gravity at small scale. The research performed by Mottola and Mazur suggests that when a star collapses due to

The Hidden Factor

gravitational instability, it would lead to the formation of a strange new object called the Gravistar in the form of a condensate bubble surrounded by a thin spherical shell composed of gravitational energy situated where the event horizon would be in a postulated black hole. According to their model, the condensate inside the shell would exert an outward pressure, which would keep it from collapsing on to it to form a black hole.

It is to be pointed out that the model of the gravitational instability proposed by Motolla and Mazur is consistent in some ways to the predictions of GNM. The Gravitational Instability Radius, R_{cg}, predicted by GNM equation (7-3) represents the shell radius of the proposed Gravistar model. For a Gravistar of 50 solar masses, both the models predict a Gravitational Stability Radius of about 150 to 200 kilometers as shown in Figure 7-5. In case of GNM, for values of the radius less than the R_{cg}, the mass energy can exist only in the form of the gravitational potential energy. If additional mass falls into the Gravistar, then for a given R_{cg}, the in-falling mass is transformed to gravitational potential energy and radiated outward in a similar manner as described in the Gravistar model by Motolla and Mazur. Both models eliminate any singularities associated with black holes, there is no event horizon to imprison light and matter, and this may explain the gamma-ray bursts that are observed in the skies. The expansion energy inside the Gravistar represents GNM Cosmological Anti-gravity Parameter described above by equation (7-11).

One feature not explained by the Motolla and Mazur model is the existence of matter inside the universe, if it is considered to be a large Gravistar. GNM, however, predicts the amount of matter inside the universe as a function of its radius or age as described in Chapter 5. GNM model can also explain much lower entropy than the black hole models, since there is no entropy generation in GNM because of absence of any energy transfer across the surface boundary of the collapsing star. As discussed in chapter 3, the energy released during a spontaneous conversion of mass to energy via a spontaneous decay is used to provide gravitational or kinetic energy to the remaining (unconverted) mass of the body or particle. Since the process of spontaneous mass-energy conversion is internal to the body and absent of any energy transfer across a boundary separating the body from its environment, there is no increase in entropy during this process. Hence, the process of spontaneous decay of particles is an isentropic and reversible process. On the other hand, motion or kinetic energy of a classical (non-decaying) body caused by an external force involves energy transfer to the body from its environment is an irreversible process leading to an increase in entropy. The existing theories, including the general theory of relativity, do not account for this isentropic process of spontaneous decay and hence over predict by several orders of magnitude the increase in entropy in a black hole as described in item 2 above. While there are still unsolved issues with the proposed Gravistar model, GNM explains the observed behavior of stars and galaxies in a consistent manner.

In summary, GNM model may explain the observations without any singularities or excessive entropy predictions of the current black hole theories. GNM thus does not support existence of black holes, or dark matter as discussed in Chapter 5, and as defined in modern cosmology.

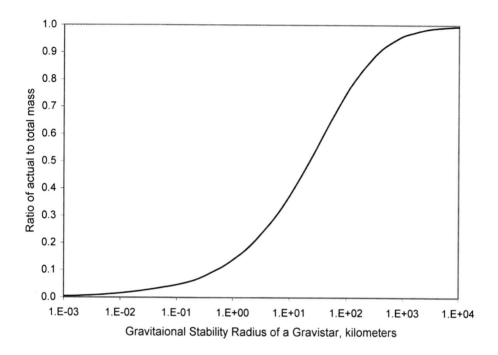

Figure 7-5: Gravitational stability predicted by GNM for a Gravistar of 50 solar masses.

Gravitational Collapse of a Bose-Einstein Condensate

As discussed in Chapter 6, when some liquids or ideal gases are cooled to sufficiently low temperatures (below 2.2 degree Kelvin) they exhibit a strange behavior called superfluidity. A superfluid exhibits no viscosity as a fraction of the atoms of the gas undergo a process called Bose-Einstein Condensation (BEC) [4]. As an ideal gas or liquid is cooled down, the motion of its individual atoms slows down and due to an increase in density the atoms begin to pack up closer together. Then, at very low temperatures close to the absolute zero degree Kelvin, the slowest moving atoms are seen to grow thousands of times from their original size overlapping each other and completely loosing their well-defined boundaries. These wraithlike atoms can pass through each other and flow against gravity. The Bose-Einstein condensate may consist of hundred of atoms which have turned into a mysterious fluid defying all classical laws of fluid mechanics and loosing its classical fluid properties. All atoms completely loose their individual identity in a so-called quantum entanglement with one another.

The conditions required for the formation of the BEC can be achieved when the particles or atoms are confined without a direct physical contact with a classical room-temperature apparatus. The condensate traps are generally very small in size, typically several micrometers in diameter. The confinement is achieved via a magnetic trap wherein reasonably high densities of atoms at extremely low temperatures can be attained by firing suitably designed laser beams. During the last few years, BEC has been achieved with rubidium, lithium, sodium and hydrogen atoms.

Earlier intuitive understanding led to an expectation that at low temperatures the attractive interactions among atoms would cause a collapse when the atoms are slowed down and situated within a close (atomic scale) distances to each other. This would prohibit formation of the BEC since there is no opposing force to the attractive force of gravity. Instead the condensing mixture of atoms would collapse upon itself leading to a microscopic implosion, releasing a lot of energy and formation of hot molecules. However, the following technical explanation is forwarded by the quantum theory for the formation of a stable BEC. According to the Heisenberg's uncertainty principle, when a particle is confined to a small spatial region, there exists an uncertainty in the momentum, which can not be zero. Because of this, the atom can never have a zero kinetic energy. This non-zero energy within the confined region is known as the Zero-point energy. Since most BEC are confined to magnetic traps of small diameter (a few micrometers), there exists a small outward pressure within the atoms that opposes the attractive gravitational forces. A stable BEC can form only when the outward repelling pressure is in balance with the attractive force of gravity. As the number of atoms in a confined trap of a certain fixed size increase, the spacing between the atoms decreases and the attractive gravitational force increases as the inverse square of the spacing between them. In the limit, for a certain population of atoms in the BEC, the gravitational force of attraction exceeds the repelling pressure or force causing a sudden collapse of the BEC. Such phenomenon has been observed in BEC experiments [37] and is similar to the phenomenon of the observed supernova collapse of stars.

Another key observation that is important for understanding the BEC phenomenon is that the measured speed of light in typical BEC is very small. Scientists [35] experimenting with BECs have been able to slow and spatially compress light pulses resulting in their complete localization and containment within an atomic cloud. Thus the measured speed of light is close to almost zero in cold vapor of sodium ions. It is intuitive to expect that as the population density or number of atoms in a BEC confined to a trap of several micrometers' size increases, the effective speed of light would decrease to almost zero.

Now, we would provide a GNM based argument to explain what leads to the gravitational collapse of a BEC. Equations (7-4) describes the estimated gravity stability radius predicted by GNM:

$$R_{cg} \approx \frac{GM_0}{C^2} \qquad (7\text{-}4)$$

For a population number, N, of atoms of mass M_a, this can be written as follows:

$$R_{cg} \approx \frac{GNM_a}{C^2} \qquad (7\text{-}14)$$

It is clear from the above equation that as the number of atoms in a given volume of BEC increases, the gravitational stability radius also increases proportionally. Another key point to note is that, as discussed above, the speed of light C decreases to almost zero with increasing number of atoms in the BEC. This would lead to a still larger increase in the gravitational stability radius. In the limit when the gravitational stability radius increases to a value above the diameter of the BEC, a collapse of the BEC would occur. During the BEC experiments with Lithium atoms described in reference [37], the collapse was observed to occur when the population exceeded approximately 1250 atoms. Figure 7-6 shows the gravitational radius predicted by GNM equation (7-14) for 100 and 1250 Lithium atoms. Since the actual value of the speed of light C, which is expected to be close to zero as discussed above, is not known during these tests, the impact of varying C over a range of values is shown in Figure 7-6. As discussed above, the gravitational stability radius predicted by GNM for 1250 atoms of Lithium becomes larger than the typical diameter of one micrometer of a BEC, leading to a collapse of BEC, when C decreases to a value less than approximately 10^{-13} meters per second.

As discussed above, quantum mechanics provides a qualitative explanation based on the Heisenberg's Uncertainty Principle for existence of the anti-gravity or repulsive force balancing the attractive gravitational force allowing formation of a stable BEC. GNM on the other hand provides a quantitative mechanistic and physically based equation for the anti-gravity parameter in terms of the Cosmological Constant in equation (7-13):

$$\Lambda_{ag} = \Lambda + \left(\frac{18Gm}{5R^3C^2}\right) \quad (7\text{-}13)$$

In summary, GNM provides a quantitative and physical understanding of the formation and collapse of a BEC observed in recent experiments.

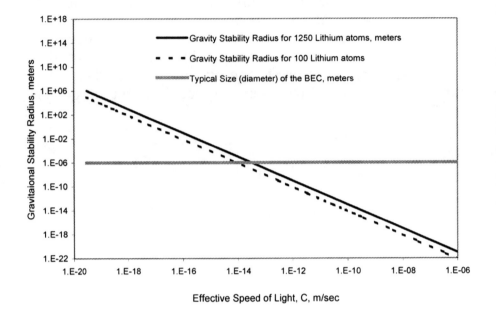

Figure 7-6: Gravity Stability Radius of Lithium atoms in a Bose –Einstein Condensate.

Matter versus Anti-matter

According to the particle theory each particle of matter has an anti-particle, which is essentially the same as the particle but with an opposite electric charge. This assumption supports the hypothesis of the Big Bang Model (BBM) that the net mass-energy of the universe is zero as discussed earlier in chapter 5. For example, a negatively charged electron has an equal but positively charged positron and a proton has an anti-proton. According to the laws of particle physics, there should be equal amounts of particles of matter and anti-matter, such that they completely annihilate each other leading to an absence of net matter in the universe. However, this is not supported by our observations of the universe, which is filled with a lot more matter in the form of stars, galaxies and clusters etc. than mostly unseen anti-matter. If there were as much or more anti-matter particles as the matter particles, the stars, galaxies and we would not be here to wonder about this paradox. Because of the missing evidence of existence of any significant amount of anti-matter, the universe seems to be biased towards the abundance of matter over anti-matter. What physical laws or mechanisms could explain such an imbalance remains an unsolved mystery of science.

We will now discuss how GNM can provide a physical perspective on existence of matter and anti-matter in the universe. As discussed earlier in this chapter, matter can exist only when there exists a balance between the gravity and anti-gravity forces or energies. If there is no anti-gravity, matter cannot exist because it would be annihilated by the gravitational pull inwards. Similarly, if there was no gravity, the anti-gravity would tear the matter apart into smaller and smaller pieces. So for a gravitationally stable mass to exist, the gravitational force or energy must be balanced by the anti-gravitational force or energy.

In chapter 5, we described GNM predictions of matter energy versus the gravitational potential and kinetic energies. The mass evolution of the universe is explained from energy considerations as depicted in Figure 5-4, which shows variation of the fractional or non-dimensional mass energy, gravitational potential energy, and kinetic energy described by equations (5-36) through (5-38). At smaller age or size, the kinetic energy is small and the universe is dominated by the gravitational potential energy that requires a substantial amount of the total maximum mass M_o to convert to the gravitational energy leading to a decreased mass of the universe. As the size of the universe increases, the gravitational energy decreases and the kinetic energy increases. At R greater than 100 billion light-years, most of the energy of the universe is in the form of kinetic energy with both the mass energy and gravitational potential energy diminishing to small values. The total fractional gravitational and kinetic energy, also defined as the fractional dark matter energy, equation (5-41), is also shown in Figure 5-4. The dark matter energy dominates at small sizes. As the size R increases, it first decreases and then increases with a minimum occurring at approximately 9 billion years which coincides with the time when the maximum universe mass occurs.

The dark matter energy of GNM can be characterized as the equivalent of the energy of the anti-matter particles in the universe. As shown in Figure 5-4, the matter energy or the existence of matter represented by the actual mass of the universe dominates only during the period of about a million light-years to 100 billion light-years. At periods earlier than this range, the gravitational potential energy annihilates the matter and during the later periods the kinetic energy breaks up the matter and diminishes its existence. Thus, according to GNM there is no need for the matter and anti-matter to be in balance during the evolution of the universe. The ratio of matter to anti-matter depends upon the size of the universe, and existence of matter is limited to only a fraction of the overall age of the universe. GNM thus provides a physical rather than a hypothetical understanding of the matter and anti-matter that is consistent with the observations of the universe.

In summary, GNM demonstrates that the classical (Newtonian) treatment of gravity, when used in conjunction with the spontaneous decay of mass to energy, properly accounts for the observed quantum and classical behavior of particles at small and large scales respectively. As such, no special or specific quantum formulation of the effects of gravity is needed.

Chapter 8

Summary and Conclusions of the Gravity Nullification Model

In spite of the great milestones achieved by science during the past few centuries, unresolved questions and unexplained paradoxes remain that threaten to pull the rug from under the picture of reality painted by the combined knowledge of science. Some scientists may proclaim victory in that the end of science has been reached and a Theory of Everything is achievable in the near future.

This work has drawn its motivation form the belief of this author that the modern science, which represents the science of the inanimate matter has reached a dead-end with no clear path to progress further. If the inanimate matter was the only and complete reality of the universe, one could envision achieving the Theory of Everything with the investigation of the matter alone. But the abundance of the evidence from scientific observations of the universe and the life in it demonstrates the existence of inherent spontaneity (or consciousness) in nature. Some examples of such spontaneity include well-accepted scientific phenomena of the wave-particle duality or spontaneous decay of unstable particles. In fact, the very existence of motion or evolution in the universe is a direct evidence of inherent spontaneity in nature. Without the spontaneity or consciousness, the inanimate matter can never experience motion at its own. Science has yet to understand the inner workings of this observed spontaneity and achieve its integration into the scientific method. Consciousness forms the fundamental basis of all scientific observations and theories, and science has only begun to investigate the question of consciousness. So far the scant treatment of spontaneity in nature has been lost in the uncertainties and cumbersome statistics of the scientific observations and theories. This author believes that the irresolvable paradoxes and the singularities in the so-called successful theories of the modern science are the direct result of the conscious exclusion and ignorance by the current scientific method of the inherent spontaneity in nature. These singularities could potentially and seriously damage the credibility and validity of the existing scientific theories. Another significant and adverse impact of this negligence and ignorance has been a great loss of simplicity, elegance and beauty in the formulations of scientific theories, which have to resort to complex, cumbersome and expensive computer solutions to obtain a useful answer.

The motivation for this work is to enhance the scientific method via inclusion of the observed spontaneity in nature into the basic formulation of physical models of the universal phenomena. This would provide, hopefully, the missing ingredient in the current scientific method to achieve its potential to represent the ultimate reality in the universe and to fulfill the human quest for the ultimate Theory of Everything.

In chapters 1 through 7, we identified and described several of the technical issues related to the current theories including the Newtonian and quantum mechanics, and theory of relativity. In order to eliminate the shortcomings in the current theories, several improvements are put forward to enhance consistency with scientific observations and improve predictability. The following is an overall summary to recapitulate the new concepts proposed in this work with some additional perspective on their significance.

Speed of Light

The second postulate of Einstein's specific theory of relativity (ESTR) states that the speed of light C remains constant in any frame of reference irrespective of the speed of the frame. This postulates leads to the limitation on the maximum speed at which an entity or signal can travel in any frame of reference. This limitation is the fixed speed of light, which also prohibits non-locality and action at a distance in the universe. The basis for this postulate is the results of the experiments such as those by A. A. Michelson and E. W. Morley, which demonstrated the constancy of the speed of light in various frames of references with a varying degree of alignment with the earth's motion and moving at different speeds.

There is a dilemma that challenges our common sense notion of physical reality when we use the concept of velocity in conjunction with the motion of light. According to the ESTR, when V approaches C, both the elapsed distance and time dilate to zero. The velocity of light, which is the ratio of elapsed distance over time, looses its meaning from a mathematical point of view (ratio of zero distance over zero time). Also, from the point of view of physical reality it becomes incomprehensible for a photon of light to move at the speed of light in a zero space with zero time. This led to the following proposed modification to the second postulate as part of the modified specific theory of relativity (MSTR):

Modified Postulate 2

2a. Elapsed distance-time is conserved during motion of light in empty space.

2b. The maximum rate at which elapsed distance is converted to time when light moves through empty space is invariant regardless the motion of the light source or the motion of the observer.

The key difference is in the interpretation of C. In the MSTR, C is described as a universal constant of conservation for the elapsed distance-time, while in the specific theory of relativity C is described as the constant velocity of light. The proposed modified interpretation of C takes away the restriction on the constancy of the speed of light or its maximum allowed limit, which is consistent with the experimental observations and interpretations of non-locality in quantum mechanics.

Another significance of this new interpretation is that the above two laws of conservation of mass-energy and space-time together make mass-energy-space-time as one continuum for the motion of light via their correspondence to a single universal constant C, as expressed earlier in equation (2-5):

$$\frac{E}{m} = \frac{S^2}{t^2} = C^2 \qquad (2\text{-}5)$$

Another aspect of the modified definition of C as a universal constant of conservation is that the empty space can now be represented as a stand-alone fixed and independent entity in the universe in which a physical entity (a particle or a body) and its motion reside. The time dilation occurs and depends upon the velocity of motion of the entity in the fixed space. The effective velocity of the entity does not have an imposed upper limit other than the limit from the maximum energy considerations. In other words, a particle or signal may still not be able to attain a speed faster than C due the fact that it may require infinite energy to accelerate to a speed C as per equation (2-3):

$$M = \frac{M_o}{\sqrt{1 - (V/C)^2}} \qquad (2\text{-}3)$$

On the other hand, a particle that spontaneously decays to zero mass can attain infinite effective speed in a selected frame of refrence due to the dilation of mass and time to zero, as described in chapter 2 and 3, and/or attainment of an infinite wavelength as discussed in chapter 4.

Since equation (2-3) predicts varying total energy for different velocity ratios, V/C, the ESTR formulation does not conserve total or universal mass-energy in relativistic frames of references that are moving at different speeds relative to each other. Such an increase in mass-energy violates the principle of mass-energy conservation. As we discussed earlier, spatial distance dilates to zero at V=C. Thus at V=C, ESTR would also lead to an infinite mass confined to a zero volume. Such a scenario provides a singularity in ESTR, which does not seem to represent a common sense physical reality. This led us to the following new postulate in the MSTR to account for mass-energy conservation in relativistic frames of references:

Postulate 3: The total mass-energy (equal to the rest mass-energy, $E_o=M_oC^2$, in the stationary frame of reference) of a system is conserved in all moving frames of references.

Mass-energy Behavior and Spontaneous Decay of Mass or Particles

One of the key differences in the formulation of the universe model in GNM versus BBM is the assumption regarding the total energy of the universe. In BBM, the total energy is assumed to remain zero throughout the evolution of the universe to justify a flat universe. In GNM, the total energy is assumed to remain constant, $E_o=M_oC^2$, in accordance with the proposed postulate 3 of the modified specific theory of relativity (MSTR). According to this postulate, the total energy (equal to the rest mass-energy, $E_o=M_oC^2$) of a system is conserved in all moving frames of references. This has the following implications on the varying states of motion of a given system:

1. The motion or kinetic energy must come from the conversion of mass-energy within the system, if no external sources of energy exist and laws of conservation are to be satisfied.
2. No two inertial frames of references attached to two separate bodies moving at a non-zero relative velocity to each other can be independent, since an

instantaneous mass-energy transfer between the two frames of references has to occur to allow the relative motion between the two frames. This is consistent with the observed connectivity or non-locality in the universe.

3. Since the Einstein's specific theory of relativity, equation (2-3) stipulates an increase in the total energy of a classical or non-decaying mass with an increase in its velocity, there must exist a finite amount of spontaneously decaying mass to provide for the increase in the kinetic energy of the non-decaying mass. This led to the concept of existence of two types of masses- the classical or the non-decaying mass and the self-decaying mass described in chapter 3. The behavior of the self-decaying mass is consistent with the widely accepted theory of unstable particles subject to a spontaneous decay.

Wave-particle Duality and Spontaneous Decay of Mass or Particles

The observed wave-particle duality of quantum particles and existence of unstable particles subject to instant and spontaneous decay are credible evidences of the spontaneous behavior in nature at small scale. A physical model of this spontaneous behavior is represented by GNM, as discussed in chapter 3 and mathematical equations for predicting the wavelength and frequency of a particle or matter-wave are developed using GNM in chapter 4.

The model proposed by Louis de Broglie in equation (4-1) has been widely accepted and used in quantum mechanics to predict wavelength of particles. This model was based on classical motion of a body without any consideration of the relativistic effects of the specific theory of relativity. The mass of the body was assumed to remain constant irrespective of the magnitude of its velocity. The velocity itself was based on the notion of fixed or non-variant space and time coordinates as in classical Newtonian mechanics. From the relativistic models of GNM presented in Chapter 3, we know that at high speeds relativistic effects become significant and mass-energy-space-time can vary in relation to each other. In GNM based wave-particle model, relativistic effects are built into equations (4-5) to (4-14). Because of the relativistic effects, the wave-particle behaviors predicted by de Broglie and GNM models are different depending upon the mass and velocity of the particle and whether the particle is self-decaying or non-decaying.

The results of the de Broglie model at small velocities, $V \approx 0$, do not match the observed reality. The model predicts that a very large mass at $V \approx 0$, will have a large wavelength and hence will act like a wave. The large stationary masses we observe in our common experience in the world act as separated bodies with well-defined and spatially identifiable boundaries. In fact, the bigger and heavier the body is, the less wavelike it acts. This is counter to the predictions of the de Broglie model. This aspect of the de Broglie model is also counter-intuitive to the Heisenberg Uncertainty and quantum mechanics theories. Such a discrepancy is understandable since this model ignores relativistic effects in spite of the fact that it was originally developed for quantum particles, such as photons, moving at large velocities close to C. At $V=0.8C$ the predictions of de Broglie model and GNM model for a self-decaying mass are fairly close as shown in Figures 4-1 and 4-2. Figure 4-3 shows the predicted wavelengths of an electron with a rest mass of 10^{-30} kilogram using different models. As discussed earlier, a good agreement is seen between GNM self-decaying mass model and de Broglie model for V in the range of $0.7C$ and $0.9C$. At very large velocities close to C, the relativistic effects become extremely significant and the de Broglie model fails to predict the resulting large increase in the wavelength and observed non-locality in quantum experiments. These effects are successfully predicted by GNM, vindicating its credibility and physical basis.

The Hidden Factor

GNM Resolves Shortcomings of BBM

As described above, one of the key differences in the formulation of the universe model in GNM versus BBM is the assumption regarding the total energy of the universe. In BBM, the total energy is assumed to remain zero throughout the evolution of the universe to justify a flat universe. In GNM, the total energy is assumed to remain constant at the currently estimated value, $E_o=M_oC^2$, as described in equation (5-16) rewritten below in terms of E_o:

$$E_o = M_o C^2 = mC^2 \left\{ \frac{1}{\sqrt{1-\left(\frac{V}{C}\right)^2}} - 1 \right\} + \frac{3Gm^2}{5R} + mC^2 \quad (8\text{-}1)$$

The total energy of the universe is thus equal to the sum of the kinetic energy, gravitational potential energy and the mass-energy. This formulation of GNM resolves the following shortcomings of BBM without resorting to the incredible scenario of inflation and forcing the assumption of maintaining a mass density of the universe exactly equal to the critical mass density throughout the evolution of the universe:

1. Singularity in BBM initial conditions
2. Uncertainty in the historical and current values of parameters
3. Unavoidable but incredible Inflation scenario to make BBM work
4. The Horizon Problem
5. The Flatness Problem
6. The Cosmological Constant Problem
7. Puzzle of Dark Matter or Dark Energy
8. Singularity in the observed accelerated expansion conditions

9. Observation of high redshift objects with lower redshift galaxies
10. Creation or formation of matter
11. Rotational and radial velocities in galaxies
12. Mystery of dark galaxies

Heisenberg's Uncertainty (a Newtonian mind-set)

The Heisenberg's uncertainty is related to the observed dual behavior of photons and other small particles in the microscopic world that act both as particles as well as waves. For quantum particles, the wavelength can be a lot larger than the physical size of the particle leading to an uncertainty of defining the exact location and momentum simultaneously. This is known as the Heisenberg Uncertainty. This uncertainty is often presumed to occur due to the direct and unavoidable impact of the measuring device or process on the motion or spatial location of the measured entity itself, especially when the measured entity is a small microscopic particle. Light is often used to observe small particles. When a photon of light strikes a particle being observed, a fraction or all of the momentum of the photon may be transferred to the particle impacting its velocity and position in an unpredictable manner.

The fundamental assumption of the fixed and independent space and time underlie the Heisenberg's uncertainty principle. Hence, the Heisenberg's uncertainty principle is an extension of the Newtonian mechanics, which is valid only for macroscopic objects moving at speeds much lower than the speed of light. For particles with higher speeds approaching the speed of light, such as electrons and photons, the relativistic effects become significant and the assumption of fixed space and time does not hold. A serious shortcoming of the Heisenberg's uncertainty principle is that it ignores the impact of relativistic effects. Ironically, the principle provides the fundamental basis for the quantum behavior of the fast moving quantum particles, which are expected to have strong relativistic effects.

The Gravity Nullification Model (GNM), which is based on the theory of relativity, supports the notion that the wave-particle behavior is a fully deterministic and not a probabilistic or uncertain phenomenon. From GNM perspective, the root cause of the predicted uncertainty is the very assumption of a fixed space-time and ignorance of the relativistic effects. In other words, GNM demonstrates (chapter 6) that the uncertainty of the Heisenberg principle is not inherent in nature or ontological, but rather a direct result of choosing an inappropriate frame of reference (the Newtonian fixed space and time) to describe a dominantly relativistic phenomenon. When the relativistic effects and their effect on space-time are properly accounted for, as in the formulations of GNM, the uncertainty in the observed particle motion would diminish to zero without any lower limits suggested by Heisenberg.

Since Heisenberg's principle forms the fundamental basis for quantum mechanics, science needs to reconsider Einstein's remark questioning the completeness of the probabilistic treatment of particles in the quantum mechanics theory. Ignorance or non-inclusion of the relativistic effects is the missing physics in the formulation of the Heisenberg's uncertainty principle leading to the incompleteness of the quantum theory.

GNM Resolves Quantum Paradoxes

A quantum particle is an illusive entity that can appear from or disappear into nothingness or vacuum, and exhibits unexplained behavior that follows weird rules involving strange properties. The quantum behavior or properties are so far different from those of the real life objects that there appears to exist two separate worlds or universes for the ordinary real life objects versus the quantum objects. There exists a big gap in the fundamental understanding of the apparent duality that exists between the behaviors of the microscopic quantum particles and macroscopic classical objects. GNM fills in this gap in knowledge and provides a physical basis for understanding the quantum behavior. This physical basis consists of the spontaneous decay of particle mass to energy and the corresponding wave-particle complimentarity. GNM (chapter 6) provides deterministic mathematical expressions of the physical limits (Planck's mass and Planck's length) that govern transition between quantum and classical behavior. GNM thus explains the inner workings of quantum mechanics and provides a physical understanding of its following well-known and still unresolved paradoxes:

1. Deterministic and local behavior of classical objects versus probabilistic and non-local behavior of quantum particles.
2. Collapse of the wave function (objective reduction) or the measurement problem (*Schrödinger's Cat* paradox).
3. Particle spin; where does it come from?
4. Super-fluidity or quantum entanglement in the Bose – Einstein Condensate.
5. Parallel universes.

Quantum Versus Classical Gravity

It continues to remain a significant challenge to modern scientists to integrate gravity with other observed forces of nature in a consistent and seamless mathematical model. Especially, a viable and common mathematical description of gravity in quantum and relativity theories has eluded scientists for the last several decades with no convergence or success in sight. All attempts to quantise gravity in existing quantum theories run into serious mathematical problems.

One of the major predictions of the general relativity theory is the phenomenon of gravitational collapse wherein the matter, without an anti-gravity force, collapses under its self-attraction forces of gravitation to an infinitely small volume leading to a singularity of infinite density. The general theory of relativity is unable to account for any special phenomenon below Planck's scale that may eliminate this singularity. GNM eliminates this singularity and provides a consistent mathematical description (Chapter 7) of the gravitational phenomenon at both large and extremely small scales extending below Planck's size, 10^{-35} meters.

GNM demonstrates that the classical (Newtonian) treatment of gravity, when used in conjunction with the spontaneous decay of mass to energy, properly accounts for both the observed quantum and classical behavior of particles at small and large scales respectively. As such, no special or specific quantum formulation of the effects of gravity is needed. GNM thus fills in the gap in the fundamental understanding of the gravitational effects that govern the behavior of small microscopic quantum particles versus the behavior of classical macroscopic objects. Specifically, GNM provides a physical basis and understanding for the following:

- Role of gravity in determining the particle or wave-like behavior of a quantum entity.
- Role of gravity in governing the relationship between Planck's mass and Planck's length.
- Effects of gravity on the observed collapse behavior of Bose-Einstein condensates at low temperatures.
- Black Holes controversies.
- Matter versus anti-matter.

Chapter 9

What does it all Mean?
Scientific Reality Versus Existence

What does GNM tell us about the scientific method and its correlation with the physical reality? Is there an absolute physical reality that exists? Is there a corresponding Theory of Everything? These aspects are discussed below.

A Perspective on Time and Evolution

Is there a standard clock or an absolute time in the universe? BBM supports a specific instant of the beginning of time (t=0) when the universe was born and an explicit instant-by-instant evolution since the beginning to the current day and age of the universe, as described eloquently by Stephen Hawking [7] in his famous book – 'A Brief History of Time'. Perplexity about the nature of time and its perceived flow from past to now and beyond into the future has preoccupied physicists, philosophers and anthropologists. Albert Einstein based on his specific theory of relativity, commented- "The past, present and future are only illusions, even if stubborn ones." Paul Davies [38] states:

> "From the fixed past to the tangible present to the undecided future, it feels as though time flows inexorably on. But that is an illusion."

> "Nothing in known physics corresponds to the passage of time. Indeed, physicists insist that time doesn't flow at all; it merely *is*. Some philosophers argue that the very notion of the

passage of time is nonsensical and that talk of the river or flux of time is founded on a misconception."

"After all, we do not really observe the passage of time. What we actually observe is that later states of the world different from earlier states that we still remember. The fact that we remember the past, rather than the future, is an observation not of the passage of time but of the asymmetry of time. Nothing other than a conscious observer registers the flow of time."

Time is thus a phenomenon relative to the frame of reference of matter, and not a universal phenomenon that exists independent or outside of a material body. Time is also closely related to the consciousness of the observer, just as it is related to the motion or velocity in the specific theory of relativity. GNM predicted solution of the universe is based on quasi-steady states that satisfy the universe energy balance and is independent of time as an explicit universal parameter. Time in GNM is an inferred parameter simply calculated by dividing the current radius of the universe by the speed of light. Hence, in the strict sense, time does not play any role in the mathematical predictions of the evolution of the universe by the physical model of GNM. In GNM, as in the Einstein's theory of relativity, time is a relative entity that depends upon the speed of the observer. There is no theoretical basis at all to formulate or measure time evolution of the universe in an absolute or standard frame of reference of time. Since there is no simultaneity in the universe due to the lack of a standard or absolute frame of reference, it makes no sense to define a point of the beginning of time (t=0), time history of evolution, time rate of expansion or contraction, or an end point of time for the universe. Any definition or quantification of the history of time is merely a meaningless exercise from a universal point of view and it is difficult to assign a rational or mathematical meaning to it.

Another common scientific assumption is that when we look into distant space away from us, we are looking into the past. The farther the object being observed, it is regarded to represent an earlier era in the age of the universe. Hence, the farthest supernovas that have been observed are assumed to belong to the universe that was only a few billion years old. If this assumption were true, then an observer situated in these supernovas and looking towards us would observe us to be situated in the beginning era of the universe. In general, for different observers situated in different regions of the universe, there is no unique history of the time or no correlation as to the beginning or the ending of the universe. Hence, the assumption that farther distances represent earlier times in the universe is self-inconsistent and without any physical basis. Since time is not an absolute entity from the universe point of view, a history of the universe in any absolute time frame or for that matter its beginning (the Big Bang) and a potential ending (Crunch or death by a run-away expansion) are physically absurd concepts.

It is to be clarified that the absence of a unique standard time strictly applies to the universe as a whole. GNM does predict a relative time or a running clock in the frame of reference of a material body within the universe, which moves at velocities less than C. Hence, time is a relative reality in the frame of reference of the matter and not a universal reality. For this reason, time has no simultaneity in the universe, and an instant of time in one frame of reference is not the same instant of time in another frame moving at a different velocity. Similarly, evolution is a relative rather than an absolute reality. Both time and corresponding evolution in a given frame of reference represent state properties of matter in it, signifying an increasing entropy and complexity. In fact, matter has no choice (free will) except to be subject to time, evolution and increasing entropy and complexity. The Zero-point energy of the vacuum (consciousness) representing the eternal and omnipresent laws on the other hand, is not subject to a changing time and evolution. The relative realities of time and

The Hidden Factor

evolution represent only a part or subset of the one whole universal reality.

In this sense, while the concept of *Creation* (time=0) lacks adequate scientific basis, the theory of evolution suffers from its limits of applicability to matter and primates or species of lower consciousness. Just as the classical or Newtonian laws apply only to the motion at low velocities (V<<C), the classical theory of evolution is valid only for application to the behavior of matter and primates of limited consciousness and provides erroneous results when extrapolated to the phenomena of higher consciousness such as the human mind and consciousness. For example, species of higher consciousness would most probably not *fight for survival* or put their life at risk for a temporal prolongation of life, which is bound to end in increased complexity, disorder and death. Theory of evolution supports no ultimate purpose or meaning for species of higher consciousness whose goal is to achieve harmony, order and the ultimate survival (a fully dilated time) without resorting to a fight using their vicious 'Tooth & Claw' to achieve domination over others. Strictly speaking, the so-called theory of evolution does not fit the definition of a classical scientific theory in that it is not capable of predicting a definite outcome or future of species in a similar manner as the Newtonian mechanics can accurately predict the future motion of a body in space and time. The theory of evolution represents a correlation or best fit to the past history of changes in species life cycle with no apparent clue to its distant future, which is dependent upon its own free will and external environment, commonly known as adaptations.

While GNM supports the existence of time and evolution as a 'Relative Reality' of the matter within the eternal space of 'Universal Reality' of the Zero-point energy, it does establish a hierarchy of levels between the time/evolution and the Zero-point energy or consciousness. Since the Zero-point energy is above and beyond time and time resides within it, it represents the highest level of existence or reality. The Zero-point energy was omnipresent even before the beginning of time and will remain so even after the time evolves to its ultimate demise or ending. GNM thus refutes the following claims of reductionism described by Paul Davies [39]:

> "In contrast Darwinism, and reductionist science in general, takes a bottom-up approach, building higher and higher levels of complexity from simple bottom-level components (cranes). An out-and-out reductionist claims that the principles vested in the bottom-level primitive entities are the only fundamental reality, and properties such as being alive, consciousness, love, reverence, sacredness etc. are derived concepts – approximate ways of thinking about complex systems, without any basis in ultimate reality."

On the other hand, GNM provides a strong mechanistic support to the emergentist view expressed by Paul Davies [39]:

> "If we view the universe as a hierarchy of levels, then I believe we find curious linkages between levels in the way the universe is organized…. I see a clear linkage (not a *causal* connection though) between the top and bottom levels – "a looping back" that connects the human mind with the fundamental workings of the cosmos, through science. And this in turn suggests that the emergence of mind has some connection with the greater cosmic scheme, and not a quirky, incidental feature that accidentally got written into the human genome."

GNM formulation provides this linkage among the top (Zero-point energy or the Cosmological Constant) and bottom levels (gravity and mass) via a universal conservation of mass-energy-space-time. The key reason for the inability or failure of the classical or reductionist method to see the above well-ordered hierarchy in the universe is the absence of the physics of spontaneity or consciousness. Reductionist method is severely limited to predict the non-spontaneous behavior of inanimate matter, and hence its application and far-stretched extrapolation to spontaneous and conscious phenomena provides erroneous and absurd results without any physical basis at all.

A Perspective on Scientific Reality

The classical scientific method relies on observation and a corresponding mathematical model or theory that predicts the observed phenomenon. The observation requires an observer, a method of observation including the measuring device or experimental equipment and of course the observed phenomenon, which is either simulated or occurring naturally. Once the observations are made, the theoretician or the analyst proposes a mathematical model or theory that predicts the observations and at the same time, is consistent with other related previous observations and theories. Further, the credibility of the observations and theories is to be established by independent observations and calculations performed by scientists other than the original experimenter or theoretician. The more diverse are the independent observations and verifications, the more credible is the theory. A Theory of Everything in this regard is the ultimate mathematical model that would predict or correlate all diverse observations made by all scientists of all the observed phenomena in the universe.

GNM supports existence of various frames of references (with varying V/C) each representing a specific state of mass-energy-space-time that constitutes the physical phenomena observed in the respective frame of reference. These frames of references exist in parallel (relative) to each other, however they are not completely independent but related to each other via mass-energy transfer between them required to satisfy the global mass-energy conservation. Such frames of references give an illusion of the so-called 'parallel universes'. Physical experience of a phenomenon observed by different observers in various frames of references would differ due to a varying degree of mass-energy-space-time dilation caused by a varying velocity of each frame. Each of these physical experiences experienced by respective observers in their own frame of reference, represents a 'Reality', or more exactly a 'Relative Reality' specific to the respective frame of reference. Note that if various frames of references in a selected set of frames happen to have an equal velocity less than C, the relativity among the many 'Relative Realities' observed by corresponding observers would diminish leading to an apparent perception of a 'Global Reality'. From the point of view of the observer, the velocity of its frame of reference represents the degree of consciousness of the observer. The degree of relativity in observations by various observers depends upon the degree of relativity in their velocities or consciousness.

An absolute or 'Universal Reality', on the other hand, wherein the relativity among all possible observers and frames of references diminishes to zero is achieved when the velocity is equal to C. The 'Universal Reality' thus represents a state of universe wherein mass and time are completely dilated to zero, with the total energy E_o of the universe existing in an absolute and eternal space of almost infinite extent (wavelength corresponding to zero mass and V=C). This state of the universe represents an absolute state wherein no relativity, locality (spatial) or temporality (time) exists with all its energy distributed homogeneously over an undivided space in perfect communication. In this state, the observers, the observed as well as the measuring equipment are all merged into the one whole 'Universal Reality'. Since the mass is dilated to zero, this is also referred to as the state of the 'Zero-point Energy', which signifies the absolute reference state of the universe. Some also refer to this as the one whole eternal and omnipresent space of the absolute consciousness.

Can a scientific experiment involving a classical (separated) observer and classical equipment measure or observe the properties of the state of the 'Universal Reality' described

The Hidden Factor

above? The answer is no, since in this absolute reference state there are no separated entities such as the observers, the observed and the measuring equipment. This leads to the well-known dilemma as to how the classical scientific method can investigate the Zero-point energy or consciousness. This also explains the physical basis for the statement – "Only consciousness can observe consciousness". Another parallel question arises- "Can a theory that only deals with inanimate matter investigate the 'Universal Reality'?" Again, the answer is no, since all matter or mass has to dilate completely to zero in the state of the 'Universal Reality' or the Zero-point Energy. A theory of inanimate matter alone that does not allow for the matter to decay spontaneously to zero is unable to predict or represent the ultimate 'Universal Reality'. Similarly, a theory that relies on an absolute and undulated time (history of time or evolution) can never be representative of the 'Universal Reality'. This is not to say that these so-called theories of inanimate matter involving absolute time can not represent 'Relative Reality' or 'Global Reality', since both of these allow existence of undulated matter and time. Another point to emphasize here is that the 'Relative Reality' described here is not an 'Illusion' or 'Myth' propagated by some religions or philosophies. An observer in the material frame of reference would perceive the observed 'Relative Reality' as an absolute reality. However, only a fully conscious observer in the frame of reference of the Zero-point energy would notice the relativity and temporality of the 'Relative Reality' as opposed to the absoluteness and eternal nature of the one wholesome 'Universal Reality' of the Zero-point energy itself.

In summary, the so-called Theory of Everything, if defined as representing the 'Universal Reality', must account for spontaneity (or consciousness) as evidenced by the motion in the universe, self-decaying particles and wave-particle duality observed in nature. Such an enhanced scientific approach that integrates spontaneity or consciousness is referred to as the *Scienciousness* in this work. A theory of inanimate or non-decaying matter alone can never be the Theory of Everything since it represents only limited, local and temporal or 'Relative

Reality', which by the evolutionary nature of matter will always be divergent and afflicted with increasing complexity, chaos and entropy.

A Perspective on Science and Religion

Perhaps one of the most famous quotes regarding the relationship between science and religion is by Albert Einstein- "Science without religion is lame, religion without science is blind". At the surface, it may seem that science and religion have two separate turfs. Science pursues the manifested or material reality while religion searches for the un-manifested or non-material reality in the universe. This apparent duality has in general divided professional scientists and orthodox religious practitioners in two separate and often adversary camps. However, Einstein expressed a cosmic religious feeling in the observed order and comprehensibility of the universe by the human mind. This cosmic religious feeling was not at all related to his disbelief in a personal God of popular religion.

Paul Davies [39], providing an eloquent perspective on science and religion, states:

> "I too share Einstein's cosmic religious feeling, a sense that order in nature revealed by science is neither arbitrary nor absurd, and there is "something going on" in the universe, something deeply ingenious and elegant, about which words like "meaning," purpose" and "design" come to mind"

Paul Davies [39] further describes some of the strengths and shortcomings of both science and religion. While most of us are familiar with the strengths of science such as its objectivity and verifiability, we will focus the following

The Hidden Factor

discussion on what perspective GNM provides on some of the key shortcomings of science identified by Paul Davies as below:

"Science proceeds on two key assumptions: the physical universe is ordered in a rational manner, and that order is, at least in part, comprehensible to human beings...But these twin assumptions are a gigantic act of faith...Even if God can be banished from the causal chain, there remains the problem of where these laws came from, and more generally, why the universe is ordered and intelligible."

"Clearly both atheistic and theistic arguments collapse in a bewildering tangle of contradictions once we attempt to go beyond the natural world and ask what lies behind physical existence. The problem always occurs when linear reasoning is applied to explanation... Do we live in a universe that came into existence for no reason, and which consists of nothing more than a collection of mindless particles moved by blind and purposeless forces towards a pointless final state? Or is there more to it than that – something deeper, something significant? Is the universe ultimately an absurdity, or is there a solid rational ground to physical existence? In short, is the universe totally meaningless, or – as I have already expressed it – is it about something... These are the questions that science may not be able to answer."

Now let us discuss the above shortcomings of science in the light of GNM. The fundamental reason for science not being able to explain the existence of order in the universe is its ignorance of spontaneity or consciousness. The existing scientific method and theories address only the physics of the inanimate matter devoid of any spontaneity. The classical scientific method attempts to force-fit the observed reality and spontaneity in nature in terms of the behavior of the inanimate matter, which due to its inertia opposes any spontaneous motion. The classical or Newtonian laws of motion presume a body of mass to be completely separated and self-contained entity bounded in dimensions of space and time, which are assumed to be fixed and independent not only from each other but also from the body. Such a restrictive and erroneous description of physical reality, works reasonably well for everyday classical objects or phenomena that occur at low velocities (V<<C), since the errors or deviations from the implicit universal laws of conservation of mass-energy-space-time are negligibly small. The material achievements of science over the fast few decades and centuries are living proofs of the successes of the classical scientific approach.

This apparent success of the classical science has led to a common misunderstanding that these laws can be extrapolated to understand the behavior of the universe phenomena or life in it. However, at large velocities (V~C) at the universe scale, the errors (related to conservation of mass-energy-space-time) become extremely large leading to the increasing disorder, chaos and uncertainty of the predicted behavior of the classical mass or related phenomena. Since the classical or non-decaying mass is restricted or constrained from spontaneous conversion and transfer of mass-energy across its fixed boundaries in a fixed space and time, the mass-energy-space-time conservation is undermined locally in the vicinity of the classical non-decaying body. If science were able to observe and record systematically the spontaneous behavior of a self-decaying mass and appropriately account for it in its theories, many of these limitations would be eliminated and the implicit order observed in the universe and governed by the simplistic laws of conservation would not be an anomaly. Hence, the key culprit in the current science in its inability to represent universal or total reality has been nothing other than its own insistence to exclude spontaneity from its adopted approach. Unknowingly, this has led to a persistent violation, howsoever small, of the laws of conservation at the universal level by

The Hidden Factor

classical science in its treatment of the universal phenomena. GNM eliminates this deficiency in the current theories and successfully predicts an orderly universe governed by simple laws of conservation of mass-energy-space-time.

With regard to the question of the comprehensibility of the universe and its orderly laws to human mind, the answer provided by GNM is the prevailing non-locality in the universe in the reference state of the Zero-point energy. The spontaneity of self-decaying mass allows its total conversion to energy leading to a state of a non-local universe with a perfect communication among all its parts leading to one homogeneous (omnipresent) whole. GNM predicts a complete dilation of time in such a state; hence this state prevails eternally in the universe without being subject to changing time and evolution. Any entity in the universe including the consciousness of the human mind can not remain detached from this non-local wholesomeness of the universe. Since the same non-local and eternal Zero-point energy also contains the laws of the universe in it, the consciousness of the human mind and laws of nature are non-locally connected in perfect communication with each other at the deepest level. On the other hand, since it requires an infinite energy to accelerate a classical or non-decaying mass, it can never achieve non-locality. Hence, the science of the inanimate matter that is non-decaying lacks appropriate physics of non-locality and can never explain comprehensibility or intelligibility of the non-local nature and laws of the universe.

GNM thus provides a physical basis as to why both atheistic and theistic arguments of the current science of the inanimate matter alone collapse in a bewildering tangle of contradictions once we attempt to go beyond the natural classical world and ask what lies behind the classical physical existence.

Now, let us look into how GNM can address the philosophical question regarding whether the universe has a purpose. If we analyze carefully, a purpose is what is by definition remains as yet unfulfilled. In other words a purpose is the mirror image of an unfulfilled desire. An unfulfilled desire represents a separation of the desirer (observer or mass) in space and time from what is desired. Hence, in a way the purpose signifies undulated space and time, which is the dominant property of a classical non-decaying mass. Moreover, purpose is directly related to the consciousness or spontaneity. Since a classical or non-decaying mass has no spontaneity or self-motivation of its own, it can not have a purpose. For a classical mass to move towards or in accordance with achieving a purpose, an external and spontaneous force (agency) is necessary. Hence, it is the spontaneous external force that determines the purpose, if any, for the classical inanimate body. For these reasons, a theory or scientific method that addresses only inanimate matter never includes the appropriate physics to account for the purpose, which implies a spontaneity or self-motivation capability. In addition, if the external force causing the motion is a singularity, such as BBM initial conditions, any purpose that may exist is lost in the emptiness or un-knowability of these initial conditions. It should not be a surprise that such a theory (BBM as an example), no matter how advanced it may be, is paralyzed and incapable of predicting a purpose in the universe. We can conclude from the discussion above that the apparent lack of purpose from the universe as predicted by the current science, specifically BBM and particle physics, is a direct reflection of the incompleteness or the missing physics of spontaneity or consciousness from these theories.

GNM corrects this deficiency by integrating the physics of spontaneity in the model of the universe with the laws of conservation of mass-energy-space-time. This integration eliminates the errors caused by a violation of the conservation laws in the current models (including BBM) of the universe via providing a physical mechanism for mass-energy transfer from one space-time state of the universe to another. GNM stipulates the ultimate reference state of the universe as the Zero-point energy state, wherein a complete eternity and omnipresence prevails signifying a total fulfillment of the ultimate purpose that human consciousness could perceive. This is the ultimate (eternal) survival that any evolution could achieve. The classical evolution considers only the survival of the body or a mere prolongation of lives of the species as the goal, but consciousness or spontaneity has much broader survival goal that extends to eternity. The classical scientific evolution theorizes bodily growth and adaptations to prolong living duration of the species, but consciousness encompasses the ultimate transcendence in time in the form of the Zero-point energy. The classical evolution promotes competition and fight for survival to beat the clock; the *Consciousness* achieves harmony and oneness to stop the clock. While 'Tooth & Claw' are symbolic of the classic scientific evolution, 'Truth & Compassion' represent consciousness.

GNM thus provides a physically consistent basis for the ultimate purpose for eternal survival or existence in the form of the Zero-point energy or the absolute consciousness of the universe. The erroneous conclusions of the current incomplete theories of the inanimate matter including the Newtonian mechanics, BBM and particles physics, have misled us to believe that we live in a universe that came into existence for no reason, and which consists of nothing more than a collection of mindless particles moved by blind and purposeless forces towards a pointless final state. GNM demonstrates that there is more to the universe than that – something deeper, something significant. The 'Ghost in the atom' is actually the 'Host in the atom'. Without this 'Host', the matter in the universe would be completely annihilated by the crushing pull of gravity. The universe is not an absurdity and there is a solid rational ground to physical existence. The universe exists beautifully and elegantly with a purpose, harmony, simplicity and order. The human mind, powered by the eternal and omnipresent Zero-point energy, is fortunate to be blessed with a capacity to comprehend such a magnificent and elegant universe. There is no need for the so-called Anthropic principle, since it does not matter whether or not the human mind is unique in this respect as compared to the rest of the life forms in other parallel universes. In fact, the Anthropic principle discriminates humans from other possible forms of existence and hence, is inconsistent with the non-locality, eternity and omnipresence of the same Zero-point energy prevailing in the universe.

With regard to the shortcomings of religion, Paul Davies [39] makes the following points:

> "The first is that religion is story-based. In cases where religions cooked up stories about the physical universe, for example the creation myths, the concepts now seem quaint and even ridiculous."

> "The second thing that is wrong with religion in relation to the physical world is its dependence on agency. Things are supposed to happen in the world because a god or spirit wills it."

> "The third thing that strikes me about all world religions is that they are homocentric, that is, they revolve around human beings and our place in the universe... Are *Homo sapiens* so central to the great cosmic scheme of things?"

We will now describe what scientific perspective GNM can provide on the above. If we analyze carefully into the characteristics of all religions mentioned above, a few common themes that appear are as follows:

- The hypothetical existence of two separate entities called the Creator (or the God) and the Creation (the universe and all things including humans in it).

- Absoluteness of time with a beginning denoting an initiation point of creation (similar to the Big Bang), a follow on evolution or emergence, and an ending as a curse of disobeying or disbelief in the Creator.

- A collective and linear logic of the human mind (and not the universal consciousness) is the ultimate judge of the universal reality and hierarchy of everything in it. In other words, the human mind is the Supreme and stand-alone observer of the ultimate reality.

As discussed earlier in this chapter and previous chapters, GNM discounts the entire hypothesis of classical religions above. The ultimate or universal reality predicted by GNM is the eternal and omnipresent state of the Zero-point energy. This is the state of one and only one wholesome existence without any separation between the Creator and the Creation, the observer and the observed, the beginning and the end, the past, present and the future, one form of life and the others, nothing and everything, the zero and the infinite, the evil and the good, and so on. The paradoxes or confusion, created by the classical or linear logic, of a divided existence of fragmented reality (mass-energy) in fragmented space and time are eliminated by GNM formulation governed by the universal conservation of mass-energy-space-time. The classical world of inanimate objects needs external forces to initiate or guide their motion. The need for an external agency such as God to create and run this world is derived from a linear extrapolation of this human experience of the material world. GNM allows spontaneous and self-motivated motion eliminating the need for external forces or agency to induce motion or change in mass-energy-space-time state of an entity. GNM thus supports a formless, nameless, eternal and non-local (omnipresent) Zero-point energy as the fundamental agency or 'Universal Reality' that represents both the Creator and the Creation. The Zero-point energy of GNM is neither the 'god-of-the-gaps' concealed in the dark corners of the universe nor the 'Co-creator' insinuating some sort of a business partnership with the humans to create and manage day-to-day running of this universe.

Time according to GNM, as discussed earlier, is a relative entity that is a state or property of the matter alone and no more than a relative reality in the frame of reference of the matter. Time has no physical existence outside the boundaries of matter. Hence, all concepts related to the absoluteness of time such as the beginning or creation, evolution, emergence, ending etc. are "no more than illusions, howsoever persistent", of an observer in the material frame of reference. Time represents only a relative reality and not the universal reality. Time is an absurd concept when applied to the universe and its so-called evolution.

The classical science and religions suffer from a persistent disease of the linear logic of the materialistic frame of reference, which violates the universal laws of conservation when extrapolated to the universal scale. This classical approach works adequately well at local or even global (earthly) scale to meet the needs of the material purposes of the *Homo sapiens*. This limited and partial success of the classical approach may have aroused a false confidence in some scientists and religious practitioners regarding its viability beyond its narrow and near-field range of applicability. This false confidence is also self-serving leading to the feeling of apparent supremacy and the homocentric notion of all world religions that the universe revolves around human beings. The extrapolation of the classical approach beyond global and even galactic scales fails miserably leading to the irresolvable dilemma of the so-called dark matter. And when extrapolated to the universe scale, the results of such approach become completely absurd violating the implicit laws, such as the laws of conservation, of the universe. Devoid of spontaneity or consciousness, science of the inanimate matter alone is blind to see the inherent order, simplicity, certainty and purpose in the universe. Similarly, devoid of consciousness and handicapped by their blind beliefs, rituals, personal gods or messiahs, world religions are unable to make their followers realize the one wholesome reality that unites all beings and existence.

In light of the above discussion, GNM can propose the following extension of the famous quote by Albert Einstein-

"Science without religion is lame, religion without science is blind... Without the consciousness both science and religion are purposeless and dead".

The Law of Certainty and Free Will

The well-accepted scientific laws of conservation of mass-energy-space-time also directly imply the existence of a 'Universal Reality', howsoever beyond the capabilities of the classical scientific method, as discussed above. The following 'Law of Certainty' is derived from these laws of conservation and stated below in the form of everyday language of a layman rather than strictly scientific terms:

"What exists now, always existed in the past and will always exist in future ('Universal Reality' is eternal). The one whole 'Universal Reality' can exist in different forms (relativity of mass-energy-space-time) with a capability of spontaneous transformation from one form to another (representing 'Relative Realities' or 'Global Realities')."

A natural corollary of the above Law of Certainty is also stated below:

"Nothing never exists, it never existed in the past nor will it ever exist in the future."

What we commonly describe as 'Nothing' or vacuum is just another form (Zero-point energy) of existence of the same total energy E_o. In fact, nothing and everything are two paradoxical states orchestrated by the linear thinking of the human mind. According to GNM, these two apparently opposite states dissolve into one whole Zero-point energy or consciousness when matter dissolves from the universe to become energy.

The laws of conservation are deterministic and not probabilistic reflecting uncertainty. Existence of inherent uncertainty in nature, as propagated by the incomplete scientific theories of the inanimate matter alone, is often used as a basis by some to support existence of free will. It is a common misunderstanding that the deterministic laws disallow existence of a free will in that they fix or predetermine the outcome of an action or event. The fact is that the natural laws are neutral to the outcome, which is totally dependent upon the choice (free will) of input or initial conditions of the event. For example, the same law of gravity helps a person walk standing up as well as cause his fall if the person tries to walk slanted or in an inappropriate manner (the choice of the walker). The heat of the same fire that cooks tasty meals to satisfy our hunger can also burn our house if not used in a disciplined manner. Hence, the existence of the universal laws allows a simultaneous co-existence of the free will. Again, determinism or absence of free will is an illusion of the classical linear extrapolation of the behavior of the inanimate matter and associated incomplete scientific theories. It will not be wrong to say that the deterministic natural laws represent the spontaneity or the free will or the consciousness itself. GNM demonstrates that the natural laws of conservation allow the universe to co-exist beautifully, elegantly, eternally and in complete harmony with the spontaneity of the free will without requiring any direct intervention or enforcement by an external agency.

The Law of Certainty supports Einstein's assertion- "God does not play dice with the universe". It says that the relative realities could be many, but the ultimate or absolute universal 'Truth is One and the only One'.

References

1. Physics for Scientists and Engineers with Modern Physics, Douglas C. Giancoli, Third Edition, Prentice Hall, 2000.
2. The Ghost in the Atom, edited by P. Davies and J. Brown, Cambridge University Press, 1997.
3. Quantum Reality - Beyond the New Physics, Nick Herbert, Anchor Books, 1987.
4. The New Physics, edited by Paul Davies, Cambridge University Press, 1998.
5. The runaway universe – the race to find the future of the cosmos, Donald Goldsmith, Perseus Publishing, 2000.
6. About Time – Einstein's Unfinished Revolution, P. Davies, Simon and Schuster, 1995.
7. A Brief History of Time, Stephen Hawking, Bantam Books, 1998.
8. Relativity – The Special and The General Theory, Albert Einstein, Three Rivers Press, 1961.
9. Was Einstein Right? – Putting General Relativity to the Test, Clifford M. Will, BasicBooks, 1993.
10. The End of Physics – The Myth of a Unified Theory, David Lindley, BasicBooks, 1993.
11. Big Bang riddles and their revelations, J. Magueijo and K. Baskerville, arXiv:astro-ph/9905393, 31 May 1999.
12. Universe in Balance, New Scientist, 16 December 2000, pp. 26-29.
13. A Measurement of the Cosmological Mass Density from Clustering in the 2dF Galaxy Redshift Survey, Peacock et.al., Nature, Vol.410, 8March 2001, pp. 169-173.
14. Cosmological Constant Problems and Their Solutions, Alexander Vilenkin, arXiv:hep-th/0106083 v2, 21 june 2001.
15. Type Ia Supernovae, the Hubble Constant, the Cosmological Constant, and the age of the Universe, Tonry, John L., arXiv:astro-ph/0105413 v2, 30 May 2001.
16. 'Seeing Red – Redshifts, Cosmology and Academic Science', Halton Arp, 1998.

17. The Rotation of Spiral Galaxies, Vera C. Rubin, Science, Volume 220, Number 4604, 24 June 1983, pp. 1339-1344.
18. Radial Distribution of the Mass-Luminosity Ratio in Spiral Galaxies and Massive Dark Cores, Tsutomu TAKAMIYA & Yoshiaki SOFUE, arXiv:astro-ph/9912567 v1, 31 Dec. 1999.
19. Central Rotation Curves of Spiral Galaxies, SOFUE et. El., arXiv:astro-ph/9905056 v1, 6 May 1999.
20. Accurate Rotation Curves and Distribution of Dark Matter in Galaxies, Y. SOFUE, arXiv:astro-ph/9906224 v1, 14 June 1999.
21. The Rotation Curve of Spiral Galaxies and its Cosmological Implications, E. Battaner & E. Florido, arXiv:astro-ph/0010475 v1, 24 Oct. 2000.
22. Rotation Curves of Spiral Galaxies, Y. SOFUE & V. RUBIN, arXiv:astro-ph/0010594 v2, 4 Nov. 2000.
23. Supermassive Black Holes and the Evolution of Galaxies, D. Richstone et. al., arXiv:astro-ph/9810378 v1, 23 Oct. 1998.
24. A Relationship between Nuclear Black Hole Mass and Galaxy Velocity, K. Gebhardt et. al., arXiv:astro-ph/0006289 v2, 29 June 2000.
25. Supermassive Black Holes Then and Now, D. Richstone, arXiv:astro-ph/9810379 v1, 23 Oct. 1998.
26. Quintessence- The Mystery of Missing Mass in the Universe, Lawrence Krauss, Revised Edition of The Fifth Essence, Basic Books.
27. Einstein & Gödel, D. Berlinski, Discover magazine, March 2002, pp. 38-42.
28. In the beginning, Eugenie Samuel, New Scientist, 7 October 2000, p.10.
29. Completely dark galaxies: their existence, properties, and strategies for finding them, N. Trentham et. al., arXiv:astro-ph/0010545 v1, 26 Oct 2000.
30. Discovery of a galaxy cluster via weak lensing, D. Wittman et.al., arXiv:astro-ph/0104094 v3, 19 July 2001.
31. Chasing shadows, New Scientist, 21 April 2001, pp. 38-41.
32. Current high energy emission from black holes, R. D. Blandford, arXiv:astro-ph/0202264 v1, 13 Feb. 2002.

33. The Large, the Small and the Human Mind, Roger Penrose et. al., Cambridge University Press, Reprinted 1999.
34. Light Speed Reduction to 17 meters per second in an Ultracold Atomic Gas, Hau, L.V. et. al., Nature 397, 1999, pp. 594-598.
35. Observation of Coherent Optical Information Storage in an Atomic Medium using Halted Light Pulses, Liu, C. et.al., Nature 409, 2001, pp. 490-493.
36. Death Star, New Scientist, January 19, 2002, pp. 27-30.
37. Direct observation of growth and collapse of a Bose-Einstein condensate with attractive interactions, J.M. Gerton et. al., Nature, Vol. 408, 7 December 2000.
38. That Mysterious Flow, Paul Davies, Scientific American, September 2002, pp. 40-47.
39. A Cosmic Religious Feeling, Paul Davies, Science and the Spiritual Quest Boston Conference, CTNS and AAAS, October 21-23, 2001.

Index

A.A Michelson and E.W. Morly, 18, 22, 278
absolute luminosity, 199
absolute reference state, 14, 294, 295
absolute time, 6, 129, 141, 202, 203, 288, 290, 295
Absorption, 95
accelerating, xiii, xiv, 106, 111, 119, 159, 160, 161, 201, 203
action at a distance, 21, 22, 33, 63, 206, 235, 278
animated matter, xi
Anthropic, 5, 13, 113, 159, 160
Anti-gravity, 260
anti-matter, 12, 110, 112, 273, 274, 287
anti-particles, 15
Aspect, 2, 30, 32, 33, 229
background radiation, 107, 109, 154, 156
BBM, x, 104, 107, 108, 109, 110, 112, 113, 114, 116, 117, 119, 123, 125, 128, 129, 135, 141, 150, 151, 152, 154, 156, 158, 160, 161, 163, 197, 201, 202, 204, 273, 280, 283, 288, 300, 301
beginning of time, xiii, 4, 107, 108, 150, 151, 152, 163, 201, 288, 289, 291
Bell's Theorem, 2
Big Bang, xiii, 4, 6, 15, 44, 104, 105, 107, 108, 109, 110, 113, 114, 116, 141, 142, 150, 151, 153, 158, 202, 261, 273, 290, 302
Biggest Blunder, 111, 119
black hole, xii, 10, 16, 167, 168, 172, 173, 176, 177, 203, 264, 265, 266, 308
black holes, 167, 172, 173, 203, 264, 265, 266, 308
Bohr, 41, 63, 206
Bose-Einstein, 16, 17, 244, 247, 252, 268, 287, 309
chaos, 11, 12, 21, 296, 298
classical measurements, xv
classical reality, xvi, 6, 228
Coherence Parameter, 57, 58, 59
collapse, xv, 4, 6, 16, 17, 106, 229, 232, 248, 251, 268, 270, 271, 286, 287, 297, 299, 309
Collapse of the Wave Function, 228, 233
complexity, xiv, xv, 10, 11, 12, 13, 290, 291, 292, 296
complimentarity, 207, 216, 286
comprehensible, xiv, xviii, 1, 5, 8, 297
condensates, 16, 17, 252, 287
connectivity, 2, 7, 15, 33, 60, 235, 236, 281
consciousness, ix, x, xi, xiii, xiv, xv, xvi, xviii, xix, 1, 6, 8, 248, 249, 276, 289, 290, 291, 292, 293, 294, 295,

297, 299, 300, 301, 303, 304, 305, 306
consciousness of the observer, xv, xvi, 289, 294
conservation constant, 28, 57, 123
constant of conservation, 24, 278, 279
cosmic religious feeling, 296
Cosmological Anti-gravity Parameter, 262, 265
Cosmological Constant, xiii, xiv, xx, 15, 106, 110, 111, 112, 113, 119, 123, 142, 146, 152, 158, 159, 160, 161, 260, 261, 262, 271, 283, 292, 307
cosmology, ix, xiv, xix, xx, 17, 248, 251, 266
cosmos, 22, 292, 307
creation, 16, 44, 150, 163, 164, 196, 203, 302, 304
Creation, 163, 284, 291, 302, 303
Creator, 302, 303
critical density, 4, 106, 123, 158
Critical Mass Density, 123
Critical quantum rest mass, 224, 225
dark energy, xiii, xx, 11
Dark Galaxies, 203
dark matter, xiii, xiv, xx, 4, 7, 11, 14, 15, 44, 51, 125, 126, 127, 129, 135, 141, 142, 145, 151, 160, 161, 164, 165, 167, 168, 173, 179, 196, 201, 204, 205, 266, 274, 304

decaying mass, xviii, 51, 52, 53, 69, 70, 71, 81, 82, 86, 103, 156, 213, 220, 222, 238, 245, 281, 282, 300
duality, 222
Einstein, 7, 15, 18, 19, 21, 22, 27, 28, 34, 37, 41, 44, 63, 90, 91, 106, 108, 111, 113, 119, 121, 152, 156, 196, 197, 201, 206, 213, 229, 237, 244, 268, 272, 277, 281, 285, 286, 288, 289, 296, 304, 306, 307, 308
elapsed distance-time, 24
elliptical galaxies, 203
emergence, 12, 292, 302, 304
Emil Mottola, 264
Emission, 95
emptiness, ix, xi, 300
end of science, xvi, 1, 7, 276
energy balance, xvii, 46, 117, 154, 166, 174, 201, 289
entanglement, xii, xvii, 16, 30, 32, 33, 40, 102, 206, 229, 230, 238, 240, 244, 245, 246, 249, 268, 286
entropy, 10, 45, 264, 265, 266, 290, 296
ESTR, 18, 21, 22, 37, 277
eternal, xi, xix, 12, 14, 290, 291, 294, 295, 299, 300, 301, 303, 305
event horizon, 264, 265
evolutionary, xiii, xx, 7, 162, 296
existence, xi, xiii, 5, 10, 11, 12, 14, 22, 44, 91, 109, 111, 128, 141, 158, 159, 167, 173, 202, 203, 204, 205,

207, 261, 265, 266, 271,
273, 274, 276, 281, 291,
293, 295, 297, 299, 301,
302, 303, 304, 305, 306, 308
expansion, xiii, xiv, 4, 10, 12,
15, 32, 44, 51, 60, 104, 105,
106, 108, 109, 111, 112,
113, 114, 119, 121, 124,
125, 135, 141, 150, 151,
152, 153, 159, 160, 161,
163, 164, 173, 187, 188,
189, 192, 197, 201, 202,
204, 236, 260, 265, 283,
289, 290
experimental observations, xiv,
28, 279
experiments, x, xii, xv, xviii, 2,
14, 17, 18, 22, 23, 30, 32,
33, 41, 64, 102, 128, 217,
229, 235, 248, 269, 270,
271, 278, 282
explosion, xiii, 255, 264
flat universe, 106, 110, 111,
117, 119, 158, 159, 280, 283
free will, x, 290, 291, 306
Freeman Dyson, 206
frequency, 61, 65, 66, 67, 68,
69, 85, 90, 91, 92, 94, 103,
196, 237, 259, 281
galactic mass distribution, 179
galactic radial velocity, 188
galaxy, 16, 104, 112, 143, 159,
162, 163, 164, 165, 166,
167, 168, 169, 173, 174,
175, 176, 177, 178, 179,
180, 181, 182, 183, 184,
187, 188, 189, 190, 191,
192, 193, 194, 195, 196,
203, 204, 308

Galileian, 18, 19, 20
gaps, xiii, 7, 21, 42, 104, 196,
303
General Relativity, 307
General Theory of Relativity,
xii, 91, 252, 286
Ghost, x, xvi, 301, 307
Ghost in the Atom, x, xvi
Global Realities, 305
God does not play dice, 1, 41,
306
GÖdel, 197, 198, 308
God-particle, 240
Gravistar, 265, 267
Gravitational Collapse, 268
gravitational instability, 254,
255, 264, 265
gravitational potential energy,
116, 117, 123, 126, 129,
151, 159, 161, 164, 174,
178, 179, 184, 216, 217,
231, 240, 253, 261, 265,
274, 283
Gravitational stability, 256,
257, 258, 267
Gravitational Stability Radius,
254, 265
Gravity, 3, 15, 40, 41, 42, 43,
46, 47, 48, 49, 51, 61, 63,
103, 104, 111, 114, 206,
210, 237, 251, 254, 260,
272, 276, 285, 286
Gravity Nullification Model,
15, 41, 42, 43, 46, 47, 48,
49, 51, 61, 63, 103, 104,
114, 206, 210, 237, 251,
276, 285
Halo, 173

313

Heisenberg, xvi, 3, 13, 16, 41,
  63, 82, 206, 207, 208, 209,
  210, 211, 212, 213, 214,
  215, 269, 271, 282, 284, 285
Heisenberg's uncertainty, 13,
  209, 284, 285
Hidden Factor, i, xii, 7, 21,
  41, 42
holistic, xviii, xx, 2, 17, 38, 46,
  104
homogeneity, xiii, 109, 154,
  156
Horizon Problem, 109, 153,
  283
Host in the Atom, xvi
Hubble, xiii, 4, 104, 105, 109,
  111, 114, 120, 121, 123,
  128, 141, 142, 147, 148,
  151, 152, 161, 163, 172,
  188, 189, 190, 240, 242, 307
Hubble Constant, 123, 128,
  141, 142, 147, 148, 163,
  240, 307
Hubble Velocity, 120
human mind, xiv, 5, 6, 291,
  292, 296, 299, 301, 303, 305
human observer, xv, 2, 3, 5, 6,
  22
Huygens, 47, 50
ideal observer, xvi
inanimate, ix, x, xi, xii, xiv, xv,
  xvi, 10, 11, 13, 276, 293,
  295, 297, 299, 300, 301,
  303, 304, 306
inanimate matter, ix, x, xi, xiii,
  xiv, xv, xvi, 10, 11, 13, 276,
  293, 295, 297, 299, 300,
  301, 304, 306

inertial frames, xvii, 18, 27,
  37, 40, 60, 236, 280
inflation, xiii, 16, 113, 119,
  150, 152, 153, 154, 156,
  158, 160, 283
Inflationary scenario, 109,
  283
initial conditions, 104, 107,
  108, 150, 151, 152, 160,
  283, 300, 306
inner workings, xi, xiv, 1, 16,
  40, 206, 276, 286
Large Scale Structures, 143
Law of Certainty, 305, 306
laws of conservation, xvii, xix,
  12, 14, 15, 16, 28, 34, 40,
  46, 158, 160, 279, 280, 298,
  300, 304, 305, 306
Limiting of Quantum
  Behavior, 216
limiting visibility radius, 175
Linear Hubble Model, 121,
  141, 144
Lithium, 245, 247, 270, 272
locality, xii, 14, 26, 34, 60, 61,
  62, 63, 102, 103, 206, 229,
  235, 237, 238, 240, 294, 299
Lorentz, 24, 25
Louis de Broglie, 64, 65, 281
luminosity, 308
luminous edge, 169, 197
luminous matter, 173, 187
luminous radiant energy, 174,
  178, 180, 187
magnetic trap, 245, 268, 269
mass to luminosity ratio, 173,
  179, 180
material reality, xvii, xviii, 296
matter ejection, 187, 196

Measurement Problem, 229, 230
Milky Way, 177, 181, 182
Modified Specific Theory of Relativity, xi
Mottola and Mazur, 264
MSTR, 18, 24, 34, 38, 278, 280
multiple dimensions, xiv, 11
Newtonian, x, xii, xvii, xviii, 13, 18, 23, 27, 30, 40, 43, 57, 60, 65, 68, 77, 104, 105, 108, 152, 173, 176, 179, 196, 208, 209, 210, 213, 236, 249, 275, 277, 282, 284, 285, 287, 291, 298, 301
non-decaying mass, 44, 51, 52, 53, 54, 56, 65, 67, 69, 70, 71, 77, 81, 82, 85, 87, 88, 89, 102, 103, 212, 215, 220, 222, 225, 227, 281, 298, 299, 300
non-locality, xii, xiii, 10, 14, 15, 16, 22, 26, 28, 34, 41, 60, 61, 62, 63, 102, 103, 156, 202, 206, 229, 232, 235, 236, 237, 238, 239, 240, 249, 278, 279, 281, 282, 299, 302
non-material, xiii, xviii, xix, 296
Nothing, 12, 14
nuclei, 109, 162
Objective Reduction, 230, 232, 234
observed behavior, xiii, 17, 40, 41, 104, 110, 152, 158, 202, 205, 266

observer, xv, xvi, 6, 7, 8, 14, 16, 18, 19, 21, 23, 24, 25, 26, 27, 37, 61, 103, 109, 153, 197, 201, 202, 209, 228, 235, 237, 248, 249, 278, 289, 290, 293, 294, 300, 303, 304
omnipresent, xix, 12, 14, 34, 290, 291, 294, 299, 301, 303
order, xi, xix, 1, 18, 21, 23, 40, 51, 68, 85, 111, 120, 129, 135, 143, 149, 156, 159, 174, 178, 197, 198, 204, 208, 221, 231, 239, 253, 254, 261, 262, 277, 291, 296, 297, 298, 301, 304
orthodox, x, xx, 12, 296
paired and opposite ejections, 163, 167
Paradox of Particle Spin, 240
parallel universes, 5, 6, 13, 15, 16, 216, 248, 249, 293, 301
particles, x, xii, xiv, xvii, xviii, xx, 2, 3, 8, 10, 11, 14, 41, 44, 45, 63, 64, 111, 116, 152, 167, 206, 207, 209, 210, 213, 222, 231, 244, 245, 246, 252, 266, 268, 273, 274, 275, 276, 281, 284, 285, 286, 287, 295, 297, 301
phase transition, 150, 163, 244, 246
photoelectric effect, 90
photon, xii, 23, 30, 33, 62, 64, 65, 90, 91, 92, 94, 95, 96, 97, 98, 99, 100, 101, 102, 103, 207, 208, 229, 237, 238, 259, 278, 284

physical reality, 23, 38, 41, 70, 203, 213, 278, 280, 288, 298
physics of the small, xi
Planck's constant, 65, 90, 208, 209
Planck's length, 219, 222, 230, 252, 253, 254, 259, 260, 262, 286, 287
Planck's Length Factor, 219, 220
Planck's mass, 129, 218, 221, 222, 232, 240, 243, 252, 253, 256, 259, 260, 261, 262, 263, 286, 287
Planck's Mass Factor, 218, 220
Podolsky and Rosen, 229
polarization, 27, 229
Powel Mazur, 264
purpose, ix, xix, 1, 11, 15, 17, 77, 291, 296, 299, 300, 301, 304, 319
quanta, 90
quantum fluctuations, 204
quantum gravity, 10, 40, 231
quantum measurement, xv, 230, 231, 241
Quantum Mechanics, 206
Quantum Paradoxes, 228, 285
quantum particles, xii, xvii, 3, 13, 14, 34, 51, 63, 70, 104, 152, 156, 207, 210, 216, 217, 252, 254, 281, 282, 284, 286, 287
quantum state reduction, 230
quasars, 162, 163
Quintessence, 112, 159, 308
redshift, 162, 163, 196, 284
Reduction of the State Vector, 228

Relative Realities, 294, 305
Relativistic Hubble Model, 120, 121, 122, 123, 141, 144, 145, 146, 147, 148, 157, 161, 239, 253
relativistic kinetic energy, 45, 115, 174, 187, 251
relativity, xii, xviii, xix, xx, 2, 3, 6, 7, 13, 15, 16, 17, 18, 19, 21, 22, 27, 37, 42, 43, 57, 91, 104, 105, 108, 111, 113, 114, 121, 152, 156, 196, 197, 201, 210, 230, 245, 249, 251, 266, 277, 278, 280, 281, 285, 286, 288, 289, 294, 295, 305
religions, xviii, 295, 302, 303, 304
Richard Feyman, 206
Roger Penrose, xi, 107, 114, 228, 229, 230, 231, 232, 309
rotation curves, 168, 169, 172, 173, 176
rotational velocities, 12, 16, 167, 170, 171, 172, 173, 174, 176, 177, 181, 187, 193, 194, 196, 240
SchrÖdinger, 228, 229, 286
science and religion, xviii, xix, 8, 296, 304, 305
Scienciousness, xvii, xviii, xix, 2, 295
self-decaying mass, 44, 45, 48, 49, 51, 52, 53, 55, 65, 66, 67, 69, 70, 71, 81, 85, 91, 96, 97, 98, 99, 100, 101, 102, 104, 210, 211, 212, 214, 217, 219, 221, 222,

224, 226, 235, 238, 239,
245, 246, 281, 282, 298, 299
simultaneity, 26, 201, 289, 290
singularities, xii, xvii, 10, 11,
13, 15, 104, 107, 120, 121,
160, 265, 266, 277
space dilation, 19, 156
space-time, xii, xvii, 3, 14, 16,
17, 23, 24, 25, 26, 29, 41,
61, 62, 63, 65, 77, 102, 150,
156, 206, 230, 232, 235,
237, 248, 249, 264, 279,
282, 285, 292, 293, 298,
300, 303, 305
Special Theory of Relativity,
xii
speed of light, xii, xviii, xx, 2,
3, 6, 10, 14, 15, 18, 19, 22,
23, 24, 27, 28, 30, 33, 34,
36, 90, 91, 92, 109, 113,
114, 120, 121, 123, 128,
141, 152, 153, 156, 172,
174, 189, 198, 201, 202,
210, 212, 213, 221, 222,
232, 235, 238, 241, 245,
246, 247, 269, 270, 277,
278, 279, 284, 289
spin, 16, 26, 40, 240, 241, 286
spiral galaxies, 167, 168, 170,
171, 172, 177, 185, 186,
187, 189, 197, 198, 203
spontaneity, x, xi, xiii, xiv, xvi,
xviii, 1, 8, 10, 14, 15, 17,
152, 217, 276, 277, 293,
295, 297, 298, 299, 300,
304, 306
Spontaneity Parameter, 57
spontaneous, xiii, xvi, 10, 15,
43, 44, 45, 51, 52, 91, 92,
104, 115, 152, 217, 223,
231, 266, 275, 276, 281,
286, 287, 293, 298, 300,
303, 305
spontaneous decay, 10, 15, 44,
45, 152, 217, 231, 266, 275,
276, 281, 286, 287
superfluid, 244, 268
superfluidity, 40, 244, 268
supernovas, 141, 160, 202, 290
superposition, 228, 230, 231,
232
surface luminosity, 179, 180,
185, 186
surface mass density, 179, 180,
185, 186
survival, 291, 300, 301
Theory of Everything, xi, xiv,
xvi, 1, 17, 42, 276, 277, 288,
293, 295
vacuum, xi, xiv, xvii, xx, 10,
12, 14, 28, 106, 107, 110,
111, 112, 113, 119, 150,
158, 159, 163, 216, 260,
261, 262, 285, 290, 305
vacuum energy, xiv, 10, 14,
106, 107, 110, 111, 112,
113, 119, 158, 159, 261
velocity ratio, 31, 37, 39, 52,
57, 69, 70, 116, 122, 141,
144, 157, 161, 245, 249,
253, 254, 279
visible universe, 197, 198
wave equation, 228
wave function, xv, 2, 6, 229,
232, 248, 286
wave-particle, xii, xviii, 11, 13,
14, 15, 45, 63, 64, 65, 68,
77, 92, 102, 104, 152, 206,

207, 209, 210, 216, 217, 218, 219, 220, 238, 244, 276, 281, 282, 285, 286, 295
weirdness, 5, 8, 10, 15, 216

Zero-point energy, xx, 14, 269, 290, 291, 292, 295, 299, 300, 301, 303, 305

About the Author

The author has Doctorate and Master's degrees in science and engineering from the reputed Massachusetts Institute of Technology. Earlier, he obtained his Bachelor of Engineering (Honors) degree from Birla Institute of Technology and Science in India. Professionally, he has been involved in state-of-the-art research and development in various fields related to science and engineering over the past 30 years. He has been a member of several professional organizations and societies. He has published extensively in professional journals and organized or chaired technical sessions at professional conferences. He has received the 'Best Paper Award' from the American Nuclear Society as well as several technical excellence awards from the reputed employer companies.

The current work represents results of author's past several years' efforts motivated by his personal interest in pursuing scientific search for reality, purpose and meaning to the universe and life.

Printed in the United States
89652LV00003B/60/A